Microplastics in Water and Wastewater

Second Edition

Edited by
Hrissi K. Karapanagioti and
Ioannis K. Kalavrouziotis

IWA PUBLISHING

Published by **IWA Publishing**
Republic – Export Building, 1st Floor
2 Clove Crescent
London E14 2BE, UK
Telephone: +44 (0)20 7654 5500
Fax: +44 (0)20 7654 5555
Email: publications@iwap.co.uk
Web: www.iwapublishing.com

First published 2020

British Library Cataloguing in Publication Data
A CIP catalogue record for this book is available from the British Library

ISBN: 9781789061680 (print)
ISBN: 9781789061697 (eBook)
ISBN: 9781789061703 (ePUB)

DOI: 10.2166/9781789061697

Microplastics in Water and Wastewater

Contents

About the Editors

Associate Professor Hrissi K. Karapanagioti was born in Patras, Greece. She graduated from the Department of the Environment at the University of the Aegean in 1992. In 1993, she moved to the USA to pursue graduate studies in the School of Civil Engineering and Environmental Science at the University of Oklahoma, receiving her Master of Environmental Science degree in 1995 and her Doctor of Philosophy in 2000, working on environmental reclamation of soils and environmental remediation of groundwater. During this time, she also spent 8 months in the Geology Department at the University of Tuebingen, Germany (1997–1998) and 1 month in the Department of Earth Sciences at the University of Waterloo, Canada studying field hydrogeology (in 1999).

In 2000, Professor Karapanagioti joined ICEHT, FORTH, Greece, as a Post-Doctoral Researcher on the EU project GRACOS, working on modeling pollution diffusion in the vadose zone. From 2001–2008, she worked as Adjunct Lecturer Professor in the Marine Sciences Department at the University of the Aegean, with a specialty in Marine Geochemical processes.

In 2007, she joined the Department of Chemistry at the University of Patras, initially as Lecturer Professor and then, from January 2013, as Assistant Professor of Environmental Chemistry with emphasis on liquid pollution (tenured in September 2016) and, since September 2018, as an Associate Professor. In 2012, she spent two months as a Visiting Professor in the School of Civil Engineering and Geosciences at Newcastle University, UK. Since 2007, she has also worked as an Adjunct Professor in the Waste Management Graduate Programme at the Hellenic Open University and, since 2011, as a part-time Level B Teacher Trainer for the Hellenic Pedagogical Institute.

In most years since 2003 Professor Karapanagioti has joined Expert Panels to review research proposals for the EU 6th, 7th and H2020 Framework Programme and, since 2010, she has also worked as a Research Programme Evaluator for the 7th EU Framework Programme in Brussels, Belgium. In 2017, she received her Master of Educational Studies from the Hellenic Open University, with a thesis on physical sciences teaching.

Professor Karapanagioti has received a number of national and international research grants. She has authored 73 papers/chapters in scientific journals and books, and edited 5 special issues in scientific journals and 2 books (including the current title). Additionally, she has published 4 papers in other scientific journals, 8 chapters in books, 2 chapters in an encyclopedia, 31 papers in conference proceedings, and has made more than 75 conference presentations, abstracts, invited talks, etc., with more than 1800 citations.

Associate Professor in the Department of Chemistry,
University of Patras
www.chem.upatras.gr/en/people/division-c/486-karapanagioti-hrissi
Twitter: @Hrissi_K

Professor Dr Ioannis K. Kalavrouziotis received his PhD in Environmental Geochemistry from the Department of Geology, University of Patras, Greece in 1999. From 2000–2013, he taught at the Department of Environmental and Natural Resources Management in Agrinio, University of Western Greece. From 2015–2018, he was Visiting Fellow in Environmental Biogeochemistry at the University of Derby, UK.

Currently, he is a Professor of the Hellenic Open University, School of Science and Technology having been appointed Director of Education on the Wastewater Management Masters Programme on 1 July 2013. He was appointed Dean the School of Science and Technology at the Hellenic Open University on 1 September 2016.

Professor Kalavrouziotis successfully completed administrative responsibilities as Agronomist of the Ministry of Agriculture (1988–2000) and was Director of the Western Greece Region Administration from 30 January 1993 to 25 November 1993. He was also a Member of the Administrative Board of the National Agricultural Research Foundation (NAGREF) (2005–2009) and was

President of the Sector for the Management of the Messologion Lagoon (Ministry of Environment, Physical Planning and Public Works) from June 2006–January 2009. He was a member of the Administrative Board of the Geotechnical Chamber of Greece (2002–2003), and member of the Administrative Board (1991–1993, 1993–1995, 1995–1997) and Secretary General (1997) of the State Agronomists Pan-Hellenic Union (PUSA). He was a member of the Administrative Board of the Organisation for the Authentication (certification) and Supervision of Agricultural Products (OPEKEPE) representing the Geotechnical Chamber of Greece (2004–2005).

He currently serves as an editorial member of several reputable journals including *Water Reuse and Desalination* (IWA Publishing), *Environment and Pollution* (Canadian Center of Science and Education), *Journal of Environmental and Analytical Toxicology*, *International Journal of Water and Wastewater Treatment*, *Frontiers in Green and Environmental Chemistry*, and *Environments*.

He is a member of the IWA and was President of the Organising and Scientific Committee of the IWA Symposium on Water, Wastewater, and Environment: Traditions and Culture (held on 22–24 March 2014 in Patras, Greece), and is a member of the interim Management Committee of the IWA Specialist Group on Water and Wastewater in Ancient Civilizations, a member of the WATERWiki Group, and a member of the IWA Working Group on the Environment.

Professor Kalavrouziotis has published 5 books and chapters, 95 peer-reviewed full research papers in international journals, 4 papers in national Greek journals, 66 international conference papers, 50 national conference papers, and more than 100 other articles in journals and newspapers with more than 2261 citations (Google scholar).

Professor and Dean the School of Science and Technology,
Hellenic Open University
www.ioanniskalavrouziotis.gr

Preface

As experts on wastewater management (I. K. Kalavrouziotis) and on liquid pollution (H. K. Karapanagioti), we decided to collaborate in editing this book after having acknowledged three things: water is a unique resource that circulates around various environmental compartments; poor wastewater management leads to liquid pollution; whilst, conversely, sustainable wastewater management can prevent liquid pollution. It seems that marine pollution from plastics and microplastics have received increased attention during the last decade. Rivers and wastewater have been identified as major sources of land-based marine plastic pollution. Identifying sources and understanding them better can help us decide on ways to prevent pollution. Thus, the main body of this book aims to describe the transport and fate of microplastics in freshwater and wastewater. At the same time, as most people do, we also wanted to explore the possible impact on humans and on the environment, and we asked three experts to write chapters devoted to impact. Finally, we agreed that the most powerful tool for battling pollution is appropriate regulation and this led to the addition of the final chapter.

The book covers the topic of microplastics in water and wastewater. Early chapters cover introductory issues related to the growing interest of the scientific community in microplastics and the human water cycle, and points where microplastics can interact with water. Subsequent chapters examine evidence of the microplastic presence in freshwaters (in both rivers and lakes) and in

Editors: Hrissi K. Karapanagioti and Ioannis K. Kalavrouziotis
doi: 10.2166/9781789061697_xvii

freshwater biota, and explore the hazardous chemicals associated with microplastics in such systems. Another set of chapters discuss the presence of microplastics in wastewater, their sources, their transfer through wastewater treatment plants, the concentration of microplastics in effluents throughout the world, the plastic biomedia used in wastewater treatment plants, and the effect on the surrounding environment of effluent pipe. These chapters also discuss the sampling methods, sample treatment and analysis techniques used for microplastics in wastewater. Additionally, the presence of microplastics in sewage sludge and in soils irrigated with wastewater or fertilised with sludge are discussed. The possible impact of plastics and their additives on plants, microalgae and humans are reviewed and presented in a critical way. Finally, a chapter summarises all the relevant regulations and initiatives that point to the necessity of a global directive for the protection of the environment from plastic and microplastic pollution.

The topic of microplastics in freshwater systems and in wastewater has scarcely been studied and requires more attention. This book aims to bring these initial findings to the attention of a broader audience and especially to operators and managers of freshwater and wastewater systems. It will also be helpful to people already aware of the marine debris problem to understand the sources of microplastics in the oceans, from freshwater systems and wastewater treatment plants.

We would like to thank all the authors who have contributed to this book, namely: N. Arsem, M. Balcer, C. Barreau, P. Bencivengo, S. A. Carr, W. Cowger, A. Dyachenko, M. Eriksen, A. B. Gray, R. R. Hurley, K. Katsanou, S. Kordella, D. P. Korfiatis, M. Lash, M. Leotsinidis, A. L. Lusher, C. Moore, N. Mourgkogiannis, L. Nizzetto, D. Papaioannou, G. Papatheodorou, L. M. Rios Mendoza, E. Sazakli, M. Thiel, J. Thompson, M. Tunalı, C. Vogelsang, and O. Yenigün. We would also like to thank the anonymous reviewers and A. Baba, L. Borea, I. Manariotis, D. Matthopoulos, M. Matouq, A. Mehra, K. Moustakas, S. Ntougias, D. Papanastasiou, M. Saidan, A. Stasinakis, A. Tanik, G. Tayfur, and D. Venieri for their helpful comments. Finally, we would like to thank IWA Publishing staff, namely, M. Hammond, N. Cunniffe and Copyeditor A. Peterson, for their assistance and guidance during the editing of this book.

Associate Professor Dr Hrissi K. Karapanagioti
Department of Chemistry
University of Patras

Professor Dr Ioannis K. Kalavrouziotis
Dean of School of Science and Technology
Hellenic Open University

Foreword (Hideshige Takada)

Research on microplastics began with relatively large microplastics (>0.3 mm) which were found in marine environments. However, a variety of sources of microplastics in the terrestrial environment were soon recognised, such as microplastic fibers from textiles and microbeads in cosmetics. These normally pass through wastewater treatment plants (WWTPs) and the extent of their removal there is critical to understanding their inputs into riverine and marine environments. These land-based sources and processes are highly diverse and variable compared to those in marine environments and, therefore, a comparative review is important to generalise the phenomena. Chapters 3 and 4 of this book provide a comprehensive review of mechanisms and removal efficiencies of microplastics during wastewater treatment. They are efficiently (95–99%) removed by a combination of primary and secondary treatment (according to Chapter 4).

Some removed microplastics are incorporated into sewage sludge and carried through composting and digestion (Chapter 6). They are then applied to agricultural lands and, finally, introduced to marine environments via surface runoff. This process is a significant source of microplastics, especially of polyethyleneterephthalate (PET) fibers to aquatic environments. In addition, some WWTPs utilise biomedia (mm-sized plastic beads of unique shape) to facilitate wastewater treatment by biofilm (Chapter 9). Chapters 7 and 8 discuss the role of

Editors: Hrissi K. Karapanagioti and Ioannis K. Kalavrouziotis
doi: 10.2166/9781789061697_xix

WWTPs as sinks or sources of microplastics to aquatic environments. One difficulty with the study of these microplastics in freshwater environments is their small size, e.g. <0.3 mm, and their association with a complex mixture of numerous natural organic and inorganic particles which interfere with the identification and quantification of microplastics.

Solid analytical methods are necessary. Chapter 5 showcases state of the art analytical methods for microplastics and claims the importance of quality control including the use of blanks, recovery and polymer identification. According to our research experience these are important issues. In particular, travel blanks to access aerial contamination of chemical fibers during sampling and treatment in open systems should be used, because a high abundance of chemical fibers in the atmosphere has been observed. Thus, reported number of microplastic fibers in past studies should be evaluated with caution.

The scope of this book also encompasses the effects of microplastics on biota. Microalgae are directly exposed to microplastics in wastewater-impacted environments (Chapter 10), whilst plants are exposed to microplastics from compost derived from sewage sludge (Chapter 11). Because humans are exposed to microplastics via a variety of routes including through drinking water in PET bottles, inhalation of air contaminated with chemical fibers, and eating seafood, the identification of the routes and understanding their contributions is important (Chapters 1 and 13). The effects on humans are discussed in terms of chemical exposure from additives and sorbed chemicals (Chapters 2 and 12) in addition to particle toxicity (Chapter 12).

All in all, the book is recommended for researchers and policymakers in the fields of environmental chemistry, civil engineering, city planning, waste management and toxicology. Furthermore, it is also worthwhile for those who are concerned about the effects of microplastics on biota and on humans.

Hideshige Takada
Professor
International Pellet Watch Project
Laboratory of Organic Geochemistry (LOG)
Tokyo University of Agriculture and Technology
Fuchu, Tokyo 183-8509, Japan

Foreword (Peter Kershaw)

The realisation that microplastics are widespread in the environment is relatively recent. The scientific community started to take a greater interest in the topic approximately fifteen years ago. Since then, much of the focus has been on the fate and effects of microplastics in the marine environment. Rather less attention has been paid to the multiple sources of microplastics and the routes by which they enter the oceans. This book provides a timely reminder of the need to consider the wider context of microplastics' generation, their transport, fate and effects at every stage of the water cycle. The range of topics covered in this volume reflects the growing body of evidence of the pervasive nature of microplastic contamination. The field is moving on from the early studies that sought to describe the distribution and abundance of microplastics largely in marine systems to include natural freshwater systems (groundwater, lakes, rivers), drinking water and wastewater. One benefit of the book is to assemble in one place the different methods required to sample and analyse microplastics in different water bodies.

'Microplastics' is a simple descriptor of what are complex physical, chemical and biological entities. Microplastic particles have intrinsic properties, such as size, shape, polymer composition and chemicals added during production to impart desired characteristics. They may acquire additional properties such as inorganic and organic films, including viruses and pathogenic bacteria. Hydrophobic

Editors: Hrissi K. Karapanagioti and Ioannis K. Kalavrouziotis
doi: 10.2166/9781789061697_xxi

contaminant chemicals present in the surrounding water will be absorbed. These properties will influence the behaviour of microplastics in the environment, and their interaction with microalgae and the possible impact that individual particles will have on organisms. This complexity is well described in the book.

For land-based sources, once released to the environment some portion of the total will enter freshwater systems directly, in run off or via atmospheric deposition. In more developed economies, a large proportion of the wastewater generated by households and commercial enterprises will enter a wastewater infrastructure and be subject to some degree of treatment. The functioning of wastewater treatment plants and the various treatments that are employed is described. The type of treatment will determine the efficacy of particulate filtration, along with the physical, chemical, and biological characteristics of the microplastics. Except in cases of very efficient particulate removal, wastewater represents a significant source of microplastics to the oceans, either directly in the case of coastal urban centres or indirectly via river transport. In the case of direct discharge, the book describes the use of numerical models to investigate the subsequent fate of particulates in the marine environment. Wastewater systems bring microplastics into close proximity with a wide range of organic compounds, and the potential role of microplastics as a vector for human and non-human pathogens is explored.

The book concludes with a plea for a more strategic approach to tackle the problem of plastics more generally. What is certain is that this book provides a very useful addition to the evidence base, on which any future strategy will depend.

Peter Kershaw
Independent consultant
Marine environmental protection

Chapter 1

Plastics and microplastics in the human water cycle

K. Katsanou[1], H. K. Karapanagioti[2] and I. K. Kalavrouziotis[3]

[1]University of Patras, Department of Geology, Patras, Greece
[2]University of Patras, Department of Chemistry, Patras, Greece
[3]Hellenic Open University, School of Science and Technology, Patras, Greece

Keywords: Drinking water, Fibers, Groundwater, Microbeads, Rivers, Wastewater, Water treatment

1.1 INTRODUCTION

The world is continually faced by the increased complexities of water pollution and its effects. For all water systems, plastics and microplastics, along with nanoplastics, are pollutants of emerging concern (Hernandez *et al.*, 2017). In the past decades, there has been a boom in the use of plastic mainly due to its properties i.e., its durability and multipurpose utility, and it is likely that the amount of plastic will continue to steadily accumulate (Horton, 2017). The constant increase in synthetic plastic production and poor management in plastic waste have led to a tremendous increase in dumping into water bodies across the world (Raza & Khan, 2018).

The sources of microplastics are both land- and ocean-based (Hammer *et al.*, 2012). Ocean-based sources represent only 20% of the total plastic debris in the marine environment (Andrady, 2011), whilst microplastics from land-based

Editors: Hrissi K. Karapanagioti and Ioannis K. Kalavrouziotis
doi: 10.2166/9781789061697_0001

sources contribute the remaining 80% (Jambeck *et al.*, 2015). Terrestrial sources come from many different origins but mainly from personal care products, air-blasting processes, improperly disposed plastics and leachates from landfill (Cole *et al.*, 2011). Once terrestrial microplastics are released into water ecosystems, most of them are transported to oceans by rivers, while the rest remain in the freshwater environment (Browne *et al.*, 2010; Li *et al.*, 2018). Although microplastics are easily transported from the source once released into the environment, it has been found that their concentrations in sediments correlate with urbanization and human activity (Horton, 2017).

Up until recently, the plastic-derived pollution in the marine environment has been the focus and thus this topic is well studied. Although large plastics, to a certain extent, can be manually removed from the environment, most plastic pollution is mainly attributed to microplastics which are not easily removed. There is also the issue of "biodegradable" and "oxodegradable" plastics that are manufactured to contain chemical bonds that can be easily degraded (esters, ethers and amides). However, these polymers still have a non-degradable hydrocarbon base and the resultant pieces left are non-degradable fragments, i.e. microplastics (Shah *et al.*, 2008). The only kinds of plastics that are totally degradable under natural conditions are bioplastics or compostable plastics (Horton, 2017).

While the vast majority of plastic sources are land-based, much less research has been focused on the investigation of their presence in freshwater ecosystems. Today, research has shifted towards inland waters. Modern studies are focused on the sources of microplastics, their pathways to marine environments, and the potential for microplastics to affect freshwater ecosystems and human health (Eerkes-Medrano *et al.*, 2015).

This chapter is an attempt to characterize the presence, levels and potential implications of microplastics in freshwaters and wastewater, as well as identify the research gaps and future priorities. Given waste management deficiencies, microplastic pollution is an unknown component of possible impact and injury to our freshwaters and freshwater-dependent biological processes.

1.2 BASIC CHARACTERISTICS OF MICROPLASTICS

Plastics are complex polymer items produced with the addition of dyes and plasticizers, etc., which give them their specific properties i.e. flexibility, durability and heat resistance. The most commonly-used polymers – and most abundant in the environment – are polyethylene (PE), polypropylene, polyester, polyvinyl chloride (PVC) and nylon.

Microplastics are particles with all their dimensions being less than 5 mm (GESAMP, 2015). They are categorized into primary and secondary microplastics: primary microplastics are originally produced to be less than 5 mm in size, ranging down to 100 nm, while secondary microplastics result from the breakdown of larger items. They include particles of a wide range of types, shapes, color and sizes. Microplastic particles can be spherical beads, fragments, fibers or films and can be made of a variety of polymers. Particles less than 100 nm are classified as nanoplastics (Rios Mendoza *et al.*, 2018).

Primary microplastics are those that are specifically manufactured to be of a small size for a specific application. They include pre-production pellets that are used in the plastic industry to manufacture larger plastic items and microbeads i. e. tiny spheres or granules added to products such as toothpastes and face scrubs for their exfoliating properties, and cosmetics for their light-reflecting properties (Browne, 2015; Cole *et al.*, 2011). Secondary microplastics are derived by the breakdown of larger plastic items and are therefore a consequence of the degradation of manufactured products due to processes such as photo-degradation, and physical, chemical and biological interactions (Galgani *et al.*, 2013; Thompson *et al.*, 2009). Some examples are plastic fragments from litter degradation, tire debris, microfibers from textiles and degradation products from road-marking paints, fishing nets, household items and other discarded plastic debris (Eerkes-Medrano *et al.*, 2015). Nanoplastics – manufactured either for research and medical purposes or formed by the degradation of microplastics – also pose an environmental threat (Koelmans *et al.*, 2016).

As already mentioned, when plastic particles break down, they gain new physical and chemical properties, increasing their potential toxic effect on organisms. Microplastics may have a toxic effect on human health (particle, chemical and microbial hazards) or may be associated with chemicals either due to the addition of plasticizer chemicals during their manufacture or by adsorption of chemicals from the environment (Takada & Karapanagioti, 2019).

1.3 THE HUMAN WATER CYCLE

There is a daily water cycle that is associated with human water consumption in urban and suburban areas which includes: water collection from a water body, treatment, storage, distribution, house use which turns it into wastewater, and then wastewater collection, transportation, treatment and discharge into the same or another water body.

Drinking water sources include surface water (e.g. rivers, lakes or dams), groundwater (springs or wells) and seawater. It is collected and sent to a water treatment plant. According to the water source, the treatment plant includes various water treatment processes. Examples of the most common treatment processes for surface water include coagulation, flocculation, sedimentation, sand and coal filtration, aeration and disinfection. For groundwater they include hardness removal, aeration, metal chemical precipitation and disinfection, whilst for seawater they include coagulation, sand and coal filtration, ultrafiltration, reverse osmosis, pH and taste adjustment, and disinfection. After treatment, water is kept in big tanks and it is usually distributed to individual houses through

Figure 1.1 Potential plastics and microplastics that can be released from a single house.

water pipes using gravity. In the past, water pipes were made of clay, cement, or PVC; today, most water pipes are made of blue high-density PE (West, 2014).

Water uses in homes include drinking, washing dishes, washing clothes, personal hygiene, house cleaning, cooking and toilet flushing (see Figure 1.1). These activities turn drinking water into wastewater, including the addition of dissolved and particulate organic matter, suspended solids, microbes, dissolved salts that increase conductivity, nutrients, surfactants and micropollutants such as caffeine, antibiotics, cosmetics, pesticides and disinfectants, etc. Wastewater is collected from each house and, by gravity, it flows towards the nearby wastewater treatment plant (WWTP) using water pipes (West, 2014).

Municipal WWTPs are expected to lower the concentration of suspended solids and dissolved organic matter in the water. In some cases, nutrient and microbe removal is required by national or local regulations. Examples of the most common treatment processes include screening, coagulation, flocculation, sedimentation, disinfection, aeration and biological treatment (through one of the following: activated sludge, biological filters, membrane biological reactors, or suspended biocarriers and through anaerobic digestion). After treatment is finalized, the effluent is sent via pipeline to the receiver which is usually a water body such as a river, a creek, a lake or the sea.

1.4 ACCUMULATION AND TRANSPORT OF MICROPLASTICS IN THE WATER CYCLE

Throughout the water cycle, there are several points where microplastics can be introduced. Water treatment and distribution can introduce microplastics into drinking water through contact with plastic machine accessories, membranes, tanks and pipes. Water use at home can introduce plastics and microplastics through personal hygiene, toilet flushing, clothes washing, etc. (see Figure 1.1).

This can happen intentionally (e.g. throwing things away down the toilet; Mourgkogiannis *et al.*, 2018), accidentally (e.g. something falling into the toilet by accident), passively (e.g. through a washing machine while washing synthetic clothes, or through the feces of humans who have accidentally ingested plastic). During wastewater treatment, microplastics can be introduced by pipes, equipment, biofilters or biocarrier media, by membranes and tanks (Karapanagioti, 2017).

Wastewater discharge pipes, WWTPs overflow systems and rivers act as the primary conduits through which plastics will travel from land to sea. It is quite certain that a large proportion of the plastics in the ocean will have passed through river systems at some point (Miller *et al.*, 2017). It is probable that during this journey some of the plastic will be retained, either temporarily or indefinitely, within the freshwater environment (Mourgkogiannis *et al.*, 2018).

A recently identified route of microplastic transport to the environment is that of airborne transport and subsequent deposition. This reveals the potential for microplastics, in the form of "urban dust" containing synthetic fibers and industrial particles, to be transported from their origins within homes and on the streets to the wider environment via the wind. It has been found that deposition of these particles is higher during rainfall events, implying that the particles become incorporated into the water droplets before hitting the ground and being washed into watercourses via runoff or drainage systems (Dris *et al.*, 2017).

1.4.1 Microplastics in seawater

The problem of plastic debris (which comprises 60–80% of total marine debris) has started to become the center of attention (Derriak, 2002) since it poses a threat to marine biota, not only because the plastics are consumed but also because they sorb potentially harmful toxins such as persistent organic pollutants (POPs) and metals, while leaching out additives such as phthalates and bisphenol A, both of which can cause adverse effects (Rochman *et al.*, 2013; Takada & Karapanagioti, 2019). Within the oceans, microplastics are widely dispersed by currents and by the wind, in addition to vertical depth transport (Horton, 2017).

The concentrations of suspended microplastics determine their bioavailability to low trophic organisms and then, possibly, promote the transfer of microplastics to higher trophic levels (Zhao *et al.*, 2015). Studies on fish demonstrated that microplastics and their associated toxins are bio-accumulated and cause problems such as intestinal damage and change in metabolic profiles (Li *et al.*, 2018).

1.4.2 The abundance of microplastics in freshwater ecosystems

Current freshwater studies suggest that microplastics are as abundant in significant numbers within freshwater environments (i.e. in rivers and lakes) as they are within the oceans. Studies of both freshwater and marine sediments have found very high concentrations – of thousands of particles per kilogram of sediment (Leslie *et al.*, 2017; Mathalon & Hill, 2014). In Europe, they have been recorded in Lake Geneva (Faure *et al.*, 2012), Lake Garda (Imhof *et al.*, 2013), the Tamar estuary (Sadri & Thompson, 2014), in the Thames (Morritt *et al.*, 2014) and in the Danube river in Austria (~3.2×10^{-4} microplastics/L) (Lechner *et al.*, 2014), and also in the Elbe, Mosel, Neckar and Rhine rivers in Germany (Klein *et al.*, 2015; Wagner *et al.*, 2014).

Moreover, microplastics have been recorded in freshwater studies worldwide: in the Great Lakes (Eriksen *et al.*, 2013), in Gauteng and North West Province, South Africa (~1.9 microplastics/L) (Bouwman *et al.*, 2018), in Lake Hovsgol, Mongolia (Free *et al.*, 2014), in Taihu Lake, China (3.4–26 microplastics/L) (Su *et al.*, 2016), in the Yangtze Estuary System, China (Zhao *et al.*, 2015), in the Three Gorges Dam (~4.1 microplastics/L), and at different locations in North America, such as the St. Lawrence River (Castaneda *et al.*, 2014), North Shore Channel of Chicago (Hoellein *et al.*, 2014), Los Angeles River, San Gabriel River, Coyote Creek (13 microplastics/L) (Moore *et al.*, 2011), Lakes Erie and St. Clair (Zbyszewski *et al.*, 2014), Lake Huron (Zbyszewski & Corcoran, 2011), and in Lakes Superior, Huron and Erie (Eriksen *et al.*, 2013). Detailed tables with microplastic concentrations in river, lake, or dam water or sediment can be found in review papers and reports (e.g. Bouwman *et al.*, 2018; Rios Mendoza & Balcer, 2019).

1.4.3 Microplastics in river water

Systems that are designed to manage storm and high precipitation events are potential entry routes for microplastics into rivers. This is because the runoff from roads and urban areas is often transferred directly to rivers to prevent flooding within populated areas, often bypassing treatment systems (Horton, 2017).

1.4.4 Microplastics in groundwater, and in tap and bottled water

Since, in many parts of the world, human and household consumption of water is covered by groundwater, more studies should be carried out in order to determine the factors involved, as well as the possible health implications. A recent study (Mintenig *et al.*, 2019) analyzed 40 m^3 of raw water and drinking water, both exclusively derived from groundwater, resulting in concentrations of 0–7 microplastic particles m^{-3}. The overall mean of 0.7 microplastics m^{-3} indicates a low contamination of drinking water with microplastics (>20 μm) when the source is groundwater. Another study analyzed samples from springs and wells from two karst aquifers in Illinois, USA (Panno *et al.*, 2019); all microplastics were fibers, with a maximum concentration of 15.2 particles/L and septic effluent given as a possible source.

Microplastics can also be present in tap water (Tyree & Morrison, 2017); in general, much lower particles were observed in treated water compared to in raw water. A recent study has also identified microplastics in bottled water (Mason *et al.*, 2018): 93% of the total bottles processed showed some sign of microplastic contamination. Mason *et al.* (2018) found roughly twice as many plastic particles within bottled water compared with their previous study of tap water (Kosuth *et al.*, 2018). Fragments were the most common particles (65%) within the bottled water studied. Bottled water contamination with microplastics partially came from the packaging and/or the bottling process itself since PET (the water bottle material) and PP (the bottle cap material) particles were predominant. Nevertheless, other types of polymers were also found. At the same time, fibers made up 98% of the microplastics within the tap water study (Kosuth *et al.*, 2018). These results indicate that the main source of the microplastic particulate is different in tap compared to bottled water.

1.5 LEGISLATION

Since microplastics have only recently been recognized as an environmental contaminant with the potential to cause ecological damage, it has taken time for governments to frame and enact policies and legislation for their manufacture, use and disposal (Horton, 2017).

It is estimated that by 2030 in Europe, the environmental damage attributed to plastic pollution will cost the equivalent of €22 billion. Thus, in order to protect the marine environment and avoid environmental damage, it is essential that the European Union take specific measures.

On 24 October 2018, the European Parliament voted in favour of earlier proposals from the European Commission to cut plastics waste, targeting in particular the single-use plastic litter polluting Europe's beaches and seas. New measures were proposed aimed at tackling marine litter at its source, targeting the ten plastic products most often found on beaches and in seas, as well as at abandoned fishing gear. On 18 January 2019, the European Council published an amended version of the draft Directive on the reduction of the impact of certain plastic products on the environment (also known as the "Single-Use Plastics Directive"). The amended draft directive significantly modifies the European Commission's proposed draft Strategy for Plastics in a Circular Economy of 28 May 2018 (EC, 2018).

The European Commission's green paper, "A European strategy on plastic waste in the environment" (EC, 2013), expressed special concern for microplastics within a review of waste legislation and highlighted potential mitigation strategies at source, saying that if microplastic pollution is deemed to be a risk to human health and priority species in freshwater environments, waste management regulation and enforcement may be necessary.

Despite the significant abundance of microplastics in several freshwater systems, there are no regulations concerning the levels of microplastics in freshwaters. However, there are several EU directives of indirect and direct relevance to microplastics pollution as an emerging contaminant. The European Marine Strategy Framework Directive is directly relevant since it included microplastics as an aspect to be measured. It aims to achieve good environmental status by 2020 (EU, 2008a) and requires prevention of litter inputs and reducing litter including microplastics in the marine environment. According to the same Directive, good environmental status will be achieved when the "properties and quantities of marine litter do not cause harm to the coastal environment" and it recommends monitoring to help achieve this outcome (Marine Strategy Framework Directive, 2008/56/EC). The Water Framework Directive (EU, 2000b), which requires monitoring of anthropogenic pressures and the protection of waters used for the abstraction of drinking water could be relevant but, so far, has not included microplastics as a descriptor. The European Drinking Water Directive (EU, 1998) requires protection against all sources of pollution but not explicitly microplastics.

Potential microplastic sources could be addressed under the following EU Directives: the Sewage Sludge Directive (EC, 1986), the Waste Framework Directive (EU, 2008b), the End-of-Life Vehicles Directive (EU, 2000a), the Waste Electrical and Electronic Equipment (WEEE) Directive (EU, 2012); the Packaging Directive (EU, 1994, as amended), the Landfill Directive (EU, 1999), and the Industrial Emissions Directive (EU, 2010), for which regulation currently only applies to the manufacture of polymers. This would greatly increase the protection of freshwater systems from industries involved in the production of plastics goods for which primary granules or pellets are required (Mahon *et al.*, 2014).

Today, in many countries worldwide, including in Canada, Ireland, the UK and the Netherlands, manufacture bans for cosmetics and personal care products containing microbeads are in effect, or laws have already been brought in (Defra, 2016). In the USA, *H.R. 1321* (the Microbead-Free Waters Act) (US Government, 2015) was signed on 28 December 2015 by President Barack Obama. The purpose of the law is to reduce water pollution caused by these products. It amends the Federal Food, Drug and Cosmetic Act to prohibit the manufacture and introduction or delivery for introduction into interstate commerce of rinse-off cosmetics containing intentionally-added plastic microbeads. Many other countries are already considering a similar ban. While microbeads are only a small contributor to the overall problem of microplastic pollution (around 2%), this is a first step towards tackling the input of unnecessary microplastics in the environment.

Finally, the European Commission decided to adopt a wider, more comprehensive approach including the Plastics Strategy, the Circular Economy Action Plan, and the revised Waste Framework Directive. It is expected that, by 2030, all plastics packaging placed on the EU market will either be reusable or will be able to be recycled in a cost-effective manner. Consumption of single-use plastics and fishing gear will be reduced and the intentional use of microplastics will be restricted.

1.6 CONCLUSIONS

Due to the limited availability of literature on freshwater microplastics, as well as the varied differences in the sampling methods used in most studies, a comparison between studies is difficult. Based on the available data, low to average levels were observed, while higher microplastic levels have been reported in developed countries such as in China, the US and in some European states.

The main complication currently is that studies that have been conducted up to now give variable results in terms of how they demonstrate categorically damaging effects on organisms. This may be because some organisms tested are more tolerant than others; different sizes and types of microplastics will be differently available to organisms with different physiological effects; or effects may be seen but only over chronic timescales (whereas many experimental studies are acute). For this reason, justifying the development of any legislation has so far been difficult.

Freshwater ecosystems therefore need to become a priority for research into the abundance, fate and effects of microplastics. It is also important to define both the environmental and biological factors that influence the concentrations and trends noticed. This need for evidence also extends to terrestrial systems where many microplastics will originate from, and where there is equal potential for ecological harm (Horton, 2017).

As a major portion of marine pollution is derived from riverine inputs, where pollution concentrations are potentially higher because they are nearer to the source, there has been a recent shift in focus to determine microplastic sources and prevent pollution and to investigate how microplastics may impact freshwater environments.

WWTPs are point sources of riverine and then of marine microplastics that are totally preventable and should be controlled (Karapanagioti & Kalavrouziotis, 2018). Easy changes within homes can capture most released microplastics, e.g. laundry machines should be regulated to include a fiber collection system (readily available and already in use in some countries). In general, prevention of pollution through changes in consumer behavior and in regulations may be the most effective way to reduce microplastics in the environment.

REFERENCES

Andrady A. L. (2011). Microplastics in the marine environment. *Marine Pollution Bulletin*, **62**(8), 1596–1605, 10.1016/j.marpolbul.2011.05.030.

Bouwman H., Minnaar K., Bezuidenhout C. and Verster C. (2018). WRC Report No.2610118. Report to the Water Research Commission by North West University, ISBN 978-0-6392-0005-7.

Browne M. A. (2015). Sources and pathways of microplastics to habitats. In: Marine Anthropogenic Litter, M. Bergmann, L. Gutow and M. Klages (eds.), Springer, Cham.

Browne M. A., Galloway T. S. and Thompson R. C. (2010). Spatial patterns of plastic debris along estuarine shorelines. *Environmental Science and Technology*, **44**, 3404–3409.

Castaneda R. A., Avlijas S., Simard M. A. and Ricciardi A. (2014). Microplastic pollution in St. Lawrence River sediments. *Canadian Journal of Fisheries and Aquatic Sciences*, **71**(12), 1767–1771.

Cole M., Lindeque P., Halsband C. and Galloway T. S. (2011). Microplastics as contaminants in the marine environment: a review. *Marine Pollution Bulletin*, **62**, 2588–2597.

Defra (Department for Environment, Food & Rural Affairs) (2016). Proposals to ban the use of plastic microbeads in cosmetics and personal care products in the UK and call for evidence on other sources of microplastics entering the marine environment. See: https://consult.defra.gov.uk/marine/microbead-ban-proposals/supporting_documents/Microbead%20ban_Consultation%20Document.pdf (accessed 15 March 2019).

Derriak J. G. B. (2002). The pollution of the marine environment by plastic debris: a review. *Marine Pollution Bulletin*, **44**, 842–852.

Dris R., Gasperi J., Mirande C., Mandin C., Guerrouache M., Langlois V. and Tassin B. (2017). A first overview of textile fibers, including microplastics, in indoor and outdoor environments. *Environmental Pollution*, **221**, 453–458, 10.1016/j.envpol.2016.12.013.

Eerkes-Medrano D., Thompson R. C. and Aldridge D. C. (2015). Microplastics in freshwater systems: a review of the emerging threats, identification of knowledge gaps and prioritisation of research needs. *Water Research*, **75**, 63–82.

Eriksen M., Mason S., Wilson S., Box C., Zellers A., Edwards W., Farley H. and Amato S. (2013). Microplastic pollution in the surface waters of the Laurentian Great Lakes. *Marine Pollution Bulletin*, **77**, 177–182.

European Commission (EC) (2013). Green paper A European strategy on plastic waste in the environment. See: https://eur-lex.europa.eu/legal-content/EN/TXT/PDF/?uri=CELEX:52013DC0123&from=EN (accessed 8 June 2019).

European Commission (EC) (2018). Proposal for a DIRECTIVE OF THE EUROPEAN PARLIAMENT AND OF THE COUNCIL on the reduction of the impact of certain plastic products on the environment. COM/2018/340 final – 2018/0172 (COD). See: https://eur-lex.europa.eu/legal-content/en/ALL/?uri=CELEX%3A52018PC0340 (accessed 15 May 2019).

European Council (EC) (1986). Directive 86/278/EEC on the protection of the environment, and in particular of the soil, when sewage sludge is used in agriculture. *Official Journal of the European Union OJ L*, **181**, 4.7.1986, 6–12. See: https://eur-lex.europa.eu/legal-content/EN/TXT/PDF/?uri=CELEX:31986L0278&from=EN (accessed 8 June 2019).

European Union (EU) (1994). Directive 94/62/EC of 20 December 1994 on packaging and packaging waste. *Official Journal of the European Union, OJ L*, **365**, 31.12.1994, 10–23. See: https://eur-lex.europa.eu/legal-content/EN/TXT/PDF/?uri=CELEX:31994L0062&from=EN (accessed 8 June 2019).

European Union (EU) (1998). Directive 98/83/EC on the quality of water intended for human consumption, *Official Journal of the European Union OJ L*, **330**, 5.12.1998, 32–54. See: https://eur-lex.europa.eu/legal-content/EN/TXT/PDF/?uri=CELEX:31998L0083&from=EN (accessed 8 June 2019).

European Union (EU) (1999). Directive 1999/31/EC of 26 April 1999 on the landfill of waste, *Official Journal of the European Union OJ L*, **182**, 16.7.1999, 1–19. See: https://eur-lex.europa.eu/legal-content/EN/TXT/PDF/?uri=CELEX:31999L0031&from=EN (accessed 8 June 2019).

European Union (EU) (2000a). Directive 2000/53/EC on end-of life vehicles – Commission Statements. *Official Journal of the European Union OJ L*, **269**, 21.10.2000, **34**–43. See: https://eur-lex.europa.eu/resource.html?uri=cellar:02fa83cf-bf28-4afc-8f9f-eb201bd61813.0005.02/DOC_1&format=PDF (accessed 8 June 2019).

European Union (EU) (2000b). Directive 2000/60/EC of the European Parliament and of the Council, establishing a framework for community action in the field of water policy. *Official Journal of the European Union OJ L*, **327**, 22.12.2000, 1–73. See: https://eur-lex.europa.eu/resource.html?uri=cellar:5c835afb-2ec6-4577-bdf8-756d3d694eeb.0004.02/DOC_1&format=PDF (accessed 8 June 2019).

European Union (EU) (2008a). Directive 2008/56/EC of the European Parliament and of the Council of 17 June 2008 establishing a framework for community action in the field of marine environmental policy (Marine Strategy Framework Directive). *Official Journal of the European Union. OJ L*, **164**, 25.6.2008, 19–40. See: http://data.europa.eu/eli/dir/2008/56/oj (accessed 15 May 2019).

European Union (EU) (2008b). Directive 2008/98/EC on waste and repealing certain Directives (Text with EEA relevance). *Official Journal of the European Union OJ L*, **312**, 22.11.2008, 3–30. See: https://eur-lex.europa.eu/legal-content/EN/TXT/PDF/?uri=CELEX:32008L0098&from=EN (accessed 8 June 2019).

European Union (EU) (2010). Directive 2010/75/EU on industrial emissions (integrated pollution prevention and control) Text with EEA relevance. *Official Journal of the European Union OJ L*, **334**, 17.12.2010, 17–119. See: https://eur-lex.europa.eu/legal-content/EN/TXT/PDF/?uri=CELEX:32010L0075&from=EN (accessed 8 June 2019).

European Union (EU) (2012). Directive 2012/19/EU on waste electrical and electronic equipment (WEEE) Text with EEA relevance. *Official Journal of the European Union OJ L*, **197**, 24.7.2012, 38–71. See: https://eur-lex.europa.eu/legal-content/EN/TXT/PDF/?uri=CELEX:32012L0019&from=EN (accessed 8 June 2019).

Faure F., Corbaz M., Baecher H. and de Alencastro L. (2012). Pollution due to plastics and microplastics in Lake Geneva and in the Mediterranean Sea. *Archives des Sciences*, **65**, 157–164.

Free C. M., Jensen O. P., Mason S. A., Eriksen M., Williamson N. J. and Boldgiv B. (2014). High-levels of microplastic pollution in a large, remote, mountain lake. *Marine Pollution Bulletin*, **85**, 156–163.

Galgani F., Hanke G., Werner S. and De Vrees L. (2013). Marine litter within the European Marine Strategy Framework Directive. *ICES Journal of Marine Science*, **70**(6), 1055–1064, 10.1093/icesjms/fst122.

GESAMP (2015). Chapter 3.1.2 Defining 'microplastics'. Sources, Fate and Effects of Microplastics in the Marine Environment: A Global Assessment. Chapter 3.1.2 Defining 'microplastics'. In: (IMO/FAO/UNESCO–IOC/UNIDO/WMO/IAEA/UN/UNEP/UNDP Joint Group of Experts on the Scientific Aspects of Marine Environmental Protection (GESAMP)), P. J. Kershaw (ed.), London, Rep. Stud. GESAMP No. 90, 96p, London.

Hammer J., Kraak M. H. and Parsons J. R. (2012). Plastics in the marine environment: the dark side of a modern gift. *Reviews of Environmental Contamination and Toxicology*, **220**, 1–44, 10.1007/978-1-4614-3414 6_1.

Hernandez L. M., Yousefi N. and Tufenkji N. (2017). Are there nanoparticles in your personal care products? *Environmental Science & Technology Letters*, **4**, 280–285.

Hoellein T. J., McCormick A. and Kelly J. J. (2014). Riverine microplastic: abundance and bacterial community colonization. In: Joint Aquatic Sciences Meeting. Portland, OR, USA. Abstract. See: http://www.sgmeet.com/jasm2014/viewabstract.asp?AbstractID=14455 (accessed 14 March 2019).

Horton A. (2017). Microplastics in the Freshwater Environment. A Review of Current Knowledge. FRR0027. Foundation for Water Research, Marlow, UK.

Imhof H. K., Ivleva N. P., Schmid J., Niessner R. and Laforsch C. (2013). Contamination of beach sediments of a subalpine lake with microplastic particles. *Current Biology*, **23**(19), R867–R868.

Jambeck J-R, Geyer R., Wilcox C., Siegler T. R., Perryman M., Andrady A., Narayan R. and Law K. L. (2015). Marine pollution. Plastic waste inputs from land into the ocean. *Science*, **347**(6223), 768–771.

Karapanagioti H. K. (2017). Microplastics and synthetic fibers in treated wastewater and sludge. In: Wastewater and Biosolids Management, I. K. Kalavrouziotis (ed.), IWA Publishing, London, pp. 77–88.

Karapanagioti H. K. and Kalavrouziotis I. K. (2018). Microplastics in Wastewater Treatment Plants: A totally preventable source. In: Sixth International Marine Debris Conference Proceedings, 41. See: http://internationalmarinedebrisconference.org/wp-content/uploads/2018/06/Sixth_International_Marine_Debris_Conference_Proceedings.pdf (accessed 15 March 2019).

Klein S., Worch E. and Knepper T. P. (2015). Occurrence and spatial distribution of microplastics in river shore sediments of the Rhine-main area in Germany. *Environmental science & technology*, **49**(10), 6070–6076.

Koelmans A. A., Bakir A., Burton G. A. and Janssen C. R. (2016). Microplastic as a vector for chemicals in the aquatic environment: critical review and model-supported reinterpretation of empirical studies. *Environmental science & technology*, **50**, 3315–3326.

Kosuth M., Mason S. A. and Wattenberg E. V. (2018). Anthropogenic contamination of tap water, beer, and sea salt. *PLoS ONE*, **13**, e0194970, doi: 10.1371/journal.pone.0194970.

Lechner A., Keckeis H., Lamesberger-Loisl F., Zens B., Krusch R., Tritthart M., Glas M. and Schludermann E. (2014). The Danube so colourful: a potpourri of plastic litter outnumbers fish larvae in Europe's second largest river. *Environmental pollution*, **188**, 177–181.

Leslie H. A., Brandsma S. H., van Velzen M. J. M. and Vethaak A. D. (2017). Microplastics en route: Field measurements in the Dutch river delta and Amsterdam canals, wastewater treatment plants, North Sea sediments and biota. *Environment International*, **101**, 133–142.

Li J., Liu H. and Chen J. (2018). Microplastics in freshwater systems: a review on occurrence, environmental effects, and methods for microplastics detection. *Water Research*, **137**, 362–374, 10.1016/j.watres.2017.12.056.

Mahon A. M., Officer R., Nash R. and O'Connor I. (2014). Scope, Fate, Risks and Impacts of Microplastic Pollution in Irish Freshwater Systems. (2014-HW-DS-2). EPA Final Report. EPA, [Wexford, Ireland].

Mason S. A., Welch V. G. and Neratko J. (2018). Synthetic polymer contamination in bottled water. *Frontiers in Chemistry*, **6**, Article 407.

Mathalon A. and Hill P. (2014). Microplastic fibers in the intertidal ecosystem surrounding Halifax Harbor, Nova Scotia. *Marine Pollution Bulletin*, **81**(1), 69–79, 10.1016/j.marpolbul.2014.02.018.

Miller M. E., Kroon F. J. and Motti C. A. (2017). Recovering microplastics from marine samples: a review of current practices. *Marine Pollution Bulletin*, **123**(1–2), 6–18, 10.1016/j.marpolbul.2017.08.058.

Mintenig S. M., Löder M. G. J., Primpke S. and Gerdts G. (2019). Low numbers of microplastics detected in drinking water from ground water sources. *Science of The Total Environment*, **648**, 631–635.

Moore C. J., Lattin G. L. and Zellers A. F. (2011). Quantity and type of plastic debris flowing from two urban rivers to coastal waters and beaches of Southern California. *Journal of Integrated Coastal Zone Management*, **11**(1), 65–73.

Morritt D., Stefanoudis P. V., Pearce D., Crimmen O. A. and Clark P. F. (2014). Plastic in the Thames: a river runs through it. *Marine Pollution Bulletin*, **78**(1–2), 196–200.

Mourgkogiannis N., Kalavrouziotis I. K. and Karapanagioti H. K. (2018). Questionnaire-based survey to managers of 101 wastewater treatment plants in Greece confirms their potential as plastic marine litter sources. *Marine Pollution Bulletin*, **133**, 822–827.

Panno S. V., Kelly W. R., Scott J., Zheng W., McNeish R. E., Holm N., Hoellein T. J. and Baranski E. L. (2019). Microplastic contamination in Karst Groundwater Systems. *Groundwater*, **57**, 189–196.

Raza A. and Zaki Khan Mohd Fahad. (2018). Microplastics in freshwater systems: a review on its accumulation and effects on fishes. *International Journal of Research and Analytical Reviews*, **5**(4), 128–140.

Rios Mendoza L. M. and Balcer M. (2019). Microplastics in freshwater environments: a review of quantification assessment. *Trends in Analytical Chemistry*, **113**, 402–408.

Rios Mendoza L. M., Karapanagioti H. and Ramírez Álvarez N. (2018). Micro(nanoplastics) in the marine environment: Current knowledge and gaps. *Current Opinion in Environmental Science & Health*, **1**, 47–51.

Rochman C. M., Browne M. A., Halpern B. S., Hentschel B. T., Hoh E., Karapanagioti H. K., Rios-Mendoza L. M., Takada H., Teh S. and Thompson R. C. (2013). Plastics and priority pollutants: a multiple stressor in aquatic habitats. *Environmental Science & Technology*, **47**, 2439–2440.

Sadri S. S. and Thompson R. C. (2014). On the quantity and composition of floating plastic debris entering and leaving the Tamar Estuary, Southwest England. *Marine Pollution Bulletin*, **81**(1), 55–60.

Shah A. A., Hasan F., Hameed A. and Ahmed S. (2008). Biological degradation of plastics: a comprehensive review. *Biotechnology Advances*, **26**, 246–265.

Su L., Xue Y., Li L., Yang D., Kolandhasamy P., Li D. and Shi H. (2016). Microplastics in Taihu Lake, China. *Environmental Pollution*, **216**, 711–719.

Takada H. and Karapanagioti H. K. (2019). Hazardous Chemicals Associated with Plastics in the Marine Environment. The Handbook of Environmental Chemistry 78. Springer-Verlag Berlin Heidelberg.

Thompson R. C., Moore C. J., vom Saal F. S. and Swan S. H. (2009). Plastics, the environment and human health: current consensus and future trends. *Philosophical Transactions of The Royal Society B Biological Sciences*, **364**, 2153–2166.

Tyree C. and Morrison D. (2017). Invisibles: the plastics inside us. Orb Media. See: https://orbmedia.org/stories/Invisibles_plastics (accessed 15 March 2019).

US Government (2015). H.R.1321: The Microbead-Free Waters Act of 2015. See: https://www.congress.gov/bill/114th-congress/house-bill/1321 (accessed 20 March 2019).

Wagner M., Scherer C., Alvarez-Muñoz D., Brennholt N., Bourrain X., Buchinger S., Fries E., Grobois C., Klasmeier J., Marti T., Rodriguez-Mozaz S., Urbatzka R., Vethaak A. D., Winther-Nielsen M. and Reifferscheid G. (2014). Microplastics in freshwater ecosystems: what we know and what we need to know. *Environmental Sciences Europe*, **26**, 12.

West J. (2014). Editorial: there's a perfect pipe for every water and wastewater project. *Municipal Sewer & Water*, (August Issue). See: https://www.mswmag.com/editorial/2014/08/theres_a_perfect_pipe_for_every_water_and_wastewater_project (accessed 16 March 2019).

Zbyszewski M. and Corcoran P. L. (2011). Distribution and degradation of freshwater plastic particles along the beaches of Lake Huron, Canada. *Water, Air, and Soil Pollution*, **220**, 365–372.

Zbyszewski M., Corcoran P. L. and Hockin A. (2014). Comparison of the distribution and degradation of plastic debris along shorelines of the great lakes, North America. *Journal of Great Lakes research*, **40**, 288–299.

Zhao S., Zhu L. and Li D. (2015). Microplastic in three urban estuaries, China. *Environmental Pollution*, **206**, 597–604.

Chapter 2

Association of hazardous compounds with microplastics in freshwater ecosystems

L. M. Rios Mendoza and M. Balcer

University of Wisconsin-Superior, Department of Natural Sciences, Superior, WI, USA

Keywords: Great Lakes, Persistent organic pollutants (POPs), Plastic pollution, Synthetic polymers, Toxic compounds

2.1 INTRODUCTION

Microplastics (MPs) have recently been found to be important contaminants in both marine and freshwater environments (Cole *et al.*, 2011; do Sul & Costa, 2014; Eerkes-Medrano *et al.*, 2015). MPs are particles with diameters less than 5 mm in size (GESAMP, 2015). Primary sources of MPs include plastics initially manufactured in small sizes such as virgin pellets or preproduction plastic, and microbeads or scrubbers that are used in personal care products. Secondary sources include the fragmentation of larger plastic items as a result of photodegradation and physical abrasion which produce smaller particles that can fall in the nanoscale size range.

Plastics contain a mixture of chemicals that are introduced during their manufacture, including additives, stabilizers, flame-retardants, pigments, fillers, and plasticizers (Hahladakis *et al.*, 2018) which can be released when MPs enter aquatic environments. In addition, MPs adsorb persistent organic pollutants (POPs), including polychlorinated biphenyls (PCBs), organochlorine pesticides (OCPs) such as dichlorodiphenyltrichloroethane (DDT) and polycyclic aromatic

Editors: Hrissi K. Karapanagioti and Ioannis K. Kalavrouziotis
doi: 10.2166/9781789061697_0015

hydrocarbons (PAHs) (Rios *et al.*, 2007). Laboratory analyzes have shown that MPs with adsorbed toxic compounds can be ingested and result in the transfer of these toxic compounds to the food web (e.g. Burns & Boxall, 2018). There is concern that toxic compounds from MPs can accumulate in aquatic organisms and ultimately in humans (Liebmann *et al.*, 2018).

The number of published studies on MPs in freshwater increased rapidly between 2010 and 2017 (Burns & Boxall, 2018), but MPs are still considered a topic of emerging concern because of the limited knowledge of their distribution and the negative effects that they can cause in aquatic ecosystems. Driedger *et al.* (2015) provided a summary of the MP research that has been conducted in the Laurentian Great Lakes and Eerkes-Medrano and Thompson (2018) presented a compilation of studies from rivers, lakes, and estuaries worldwide. These investigations primarily concentrated on determining the sources, types and abundance of MPs in freshwater. Other studies examined the rate of transport of MPs from rivers to estuaries and oceans (Browne *et al.*, 2010). Moore *et al.* (2011) reported the contribution of 2 billion (2×10^9) plastic particles from two rivers in California to the Pacific Ocean in a 72-h period. While riverbanks and wastewater treatment plants (WWTPs) have been identified as major sources of MPs (McCormick *et al.*, 2016; Rech *et al.*, 2014;), atmospheric sources cannot be ignored since some MPs can be transported by wind in the same manner as volatile pollutants (Lim *et al.*, 2018).

While knowledge of MP abundance and distribution in freshwater is increasing, little attention has been given to the role of MPs in transporting toxic materials to these ecosystems. Rios and Evans (2013) provided the first report of POPs being adsorbed onto microplastics from Lake Erie, but few studies have been published since that time. This chapter will examine the present state of knowledge on the ability of POPs, heavy metals, pharmaceuticals and other hazardous materials, including bacteria and viruses to be adsorbed onto MPs and enter freshwater environments.

2.2 HAZARDOUS COMPOUNDS ASSOCIATED WITH MICROPLASTICS

Plastics in general are chemically inert and resistant to degradation in the environment, but studies on plastic debris in the ocean have shown that plastics adsorb, concentrate and transport hydrophobic toxic compounds, pharmaceuticals, heavy metals, bacteria and viruses. Additionally, they can leak toxic compounds that are added during production of the plastic items. MPs are hypothesized as a vector of toxic compounds to the food web where bioaccumulation may occur and cause chemical and physical negative effects in organisms (Lagana *et al.*, 2018; Lobelle & Cunliffe, 2011; Ma *et al.*, 2016; Mato *et al.*, 2001; Rios *et al.*, 2010; Rochman *et al.*, 2013; Zettler *et al.*, 2013). The adsorption of hydrophobic toxic compounds present in the environment on the surface of MPs is challenging to assess, and few analyses of these compounds in freshwater systems have been performed.

2.2.1 POPs adsorbed on microplastic surfaces

POPs are ubiquitous organic environmental contaminants. They are chemically stable, lipophilic and tend to accumulate in the food web. POPs can travel long distances from their original source and tend to sorb to suspended organic matter, sediments and plastic debris. Some of the most common POPs include PCBs, OCPs and PAHs. PCBs are synthetic compounds consisting of up to 209 congeners that were used in electric power industries and as plasticizers until they were banned in the 1970s. PAHs are compounds formed from incomplete combustion, while OCPs are used mainly in agriculture. Although some of these compounds have also been banned, they persist in the environment. Wang *et al.* (2018) provided a review of factors affecting the sorption of chemicals to MPs and showed that MPs act as vectors for POPs, although their contribution to bioaccumulation and toxicity in aquatic life is still unknown.

MPs can adsorb hydrophobic toxic compounds because of their lipophilic surfaces. The dynamic of the sorption/desorption processes of several hydrophobic compounds to MPs have been evaluated under laboratory conditions, and the rates of equilibrium partitioning between MPs and seawater were determined (Hirai *et al.*, 2011; Lee *et al.*, 2019; Teuton *et al.*, 2007; Van *et al.*, 2012; Wang *et al.*, 2018).

Plastic debris samples from the ocean have been reported to contain POPs adsorbed to their surfaces (Bakir *et al.*, 2014; Mato *et al.*, 2001; Rios *et al.*, 2007). Hong *et al.* (2018) conducted an extensive literature survey and provided detailed tables of the concentrations of PCBs, OCPs, PAHs and other compounds contained on floating and beached marine plastics. Considerable variation in reported concentrations may be due to the type, size and shape of the plastic polymers analyzed, residence time in the ocean, location relative to sources of chemical contamination and analytical methods used by different researchers.

Information on the interactions of POPs with plastic debris in freshwater systems is much more limited (Table 2.1); most of the freshwater data is derived from research by this chapter's first author and her laboratory. The methods used for extraction and quantification of POPs were based on Rios *et al.* (2010) with minor modifications. Briefly, 1 g of MPs was Soxhlet extracted with dichloromethane for 24 h. The extract was spiked with surrogate mass spectrometry standards. A glass column packed with silica gel (5% deactivated) was used to clean the extract. Analytes and internal standards were eluted using 40 mL of a mixture of dichloromethane:hexane (25:75). The identification and quantitation of toxic compounds were achieved by gas chromatograph with a mass spectrometer (GC/MS). The identification of the types of synthetic plastic polymers was made by Fourier transform infrared spectroscopy attenuated total reflectance (FTIR-ATR).

Table 2.1 Concentrations (ng g^{-1}) of toxic compounds on plastic debris from freshwater environments.

Location	Σ PCBs	Σ PAHs	Σ OCPs	Reference
Laurentian Great Lakes	ND–575	77–812	NR	Rios and Evans (2013)
Laurentian Great Lakes	ND–9856	12000–15200	NR	Rios *et al.* (2016)
Saint Louis River/Lake Superior	ND	47–20255	ND	Rios *et al.* (2018)
Swiss Lakes	0.4–548	86–5714	1.4–2715	Faure *et al.* (2015)

NR: not reported; ND: not detected; Σ: sum of compounds.

Rios and Evans (2013, data unpublished), presented the first report of POPs adsorbed onto MPs from the Laurentian Great Lakes (21 samples). They analyzed PAHs (20 compounds) and found fingerprinting of both pyrogenic and petrogenic sources. In 2016, Rios and collaborators presented the results from a set of samples collected during summer 2014 (44 samples) and analyzed PCBs (41 congeners) and PAHs that showed fingerprinting of pyrogenic sources (data unpublished). Figure 2.1 shows the variation in the types of MPs found in one sample that was collected near a WWTP on Lake Erie. In 2018, Rios *et al.* presented data on POPs adsorbed on MPs samples (17 samples) from the St. Louis River Estuary and western Lake Superior. PAHs were quantified but PCBs and OCPs (20 compounds) were not detected at the limit of detection (data unpublished). Faure and collaborators (2015), analyzed 14 surface and six beach samples from four Swiss lakes and reported concentrations of 12 PCB congeners, 16 PAH compounds and 19 OCPs. The methodology used was based on Hirai *et al.* (2011).

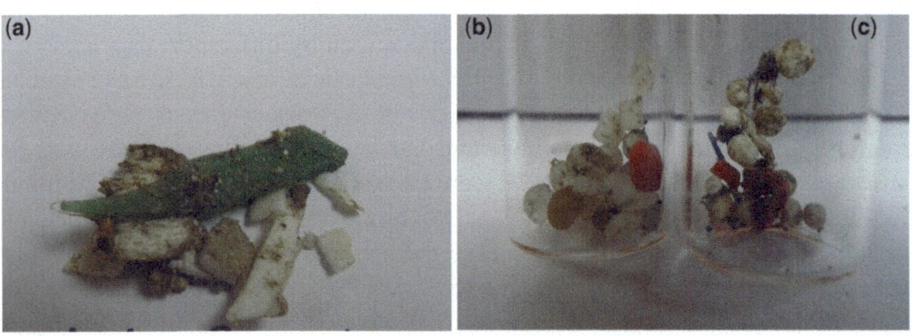

Figure 2.1 Components of one sample of MPs taken from Lake Erie, near the Cleveland Easterly Wastewater Treatment Plant, showing the variety of MP particles present: (a) fragments; (b) pre-production pellets; and (c) foam pellets, microbeads and other small fragments.

2.2.2 Pharmaceutical products

Some reports have shown that pharmaceutical products and their metabolites are toxic to organisms at low concentrations and are able to bioaccumulate in freshwater systems (Xie *et al.*, 2017; Zenker *et al.*, 2014). The areas with highest concentrations were close to wastewater discharges, mainly because waste treatment cannot completely remove these compounds.

Adsorption experiments conducted under laboratory conditions have demonstrated that MPs have the capacity to adsorb pharmaceutical compounds. Li *et al.* (2018) compared the adsorption of five antibiotics on five types of synthetic polymers in freshwater and seawater systems; the results showed higher adsorption in freshwater than in marine conditions. However, there are no reports of pharmaceutical substances adsorbed onto MPs from natural freshwater systems.

2.2.3 Metals

Metals are used as additives in some plastic products and can be leached to the aquatic environment. However, metals can also be adsorbed from aquatic environments to MPs. Studies of marine MP debris have shown that marine sediments adsorb more metals than MPs do (Ashton *et al.*, 2010; Dobaradaran *et al.*, 2018; Holmes *et al.*, 2012). The low metal concentrations in MPs could be a function of their smaller relative surface area. Mato *et al.* (2001) calculated the geometric surface area of polyethylene (PE) pellets in the range of $cm^2\ g^{-1}$ while sediments had areas in the order of $m^2\ g^{-1}$ (Millward, 1995).

2.2.4 Bacteria and viruses

MPs can also provide a place for bacteria and viruses to colonize and grow, forming biofilms which can be transported for long distances from their origin (Lobelle & Cunliffe, 2011; Reisser *et al.*, 2014; Zettler *et al.*, 2013). In the marine environment, several forms of microbes have been detected on MPs and the diversity of the bacterial community found has been correlated with various environmental factors such as season, location, substrate and age (De Tender *et al.*, 2015; Mincer *et al.*, 2016). Dang and Lovell (2000) showed that bacterial colonization was formed within 24 h and disappeared in 72 h. Studies in oceans have found MPs colonized by *Vibrio spp.*, which are human pathogens (Kirstein *et al.*, 2016; Zettler *et al.*, 2013) and *Aeromonas salmonicida* which is a fish pathogen (Viršek *et al.*, 2017). Zettler *et al.* (2013) found that bacteria on MPs were different from those in surrounding waters, suggesting that MPs are a distinct habitat for microbial communities called the Plastisphere. Similar observations were made for MPs in freshwater systems (McCormick *et al.*, 2014). However, there is little information about these communities of bacteria and viruses in lakes and rivers.

Table 2.2 Concentrations (ng g^{-1}) of plastic additives on plastic debris from freshwater environments.

Location	Σ PBDEs	BPA	Nonylphenol	Σ Phthalates	Reference
Swiss Lakes	0.2–419	4.8–28	0–612	528–111604	Faure *et al.* (2015)

Σ sum of compounds.

2.2.5 Additive compounds from MPs

During the manufacture of plastic items, a variety of chemical compounds are added to change the characteristics of the final product (Hahladakis *et al.*, 2018). Additives can comprise up to 50% of the plastic product (Hong *et al.* 2018) and can give e.g. color and resistance to heat and aging. Phthalate esters (plasticizers), bisphenol A (BPA), brominated flame retardants (BFRs), phosphorus flame retardants (PFRs), antioxidants and stabilizers are plastic additives that are known to be toxic compounds. Perflouroalkylated compounds (PFAS) are a large group of hydrophobic and lipophilic chemicals used in polymer production. Faure *et al.* (2015) reported the concentration of 14 polybrominated diphenyl ethers (PBDEs), bisphenol A (BPA), nonylphenol and seven phthalates from MPs collected from Swiss lakes (Table 2.2). Llorca *et al.* (2018) reported the adsorption rates of PFAs using natural freshwater under laboratory conditions and the results showed that PFAs had their maximum adsorption of 25% concentration of the compounds after seven day.

2.3 DISCUSSION

Plastic materials provide many benefits to society. However, the use and abuse of plastics, including inadequate waste disposal, have led scientists to note that plastics are not biodegradable or decomposable in any reasonable scale of time. Although plastics can photodegrade, this only results in producing smaller MP particles. UNEP (2016) recognized that plastic debris pollution is a real threat for humanity and our ecological environment; in fact, we are facing the "Plastic Age" and with it the addition of one more environmental issue: MPs.

One of the main sources of MPs to freshwater systems are the effluents from WWTPs; however, the importance of this source can depend on the kind of treatment used. Prata (2018) showed that tertiary treatment can retain more than 97% of MPs, but sand filters are less efficient and may be responsible for the formation of more MPs by decreasing particle size through abrasion.

MPs are vectors of toxic compounds because they accumulate persistent organic compounds, heavy metals, pharmaceutical compounds, bacteria and viruses. This accumulation is related to particle size, with smaller particles accumulating larger concentrations of toxic compounds due to their large surface–volume ratio, surface weathering and accumulation of microorganisms (biofilm). The type of polymer also affects the rate of concentration of toxic compounds with PE particles having higher adsorption rates than polypropylene (PP) particles (Wang *et al.* 2018).

There are several studies of the negative effects of ingestion of MPs by both marine and freshwater organisms under laboratory conditions (Harmon, 2018). Laboratory experiments have also examined the adsorption of toxic compounds from MPs (e.g. Chae & An, 2017). Lee *et al.* (2019) assessed the ingestion of MPs containing OCPs in artificial gut solution, their model showing fast desorption of these toxic compounds from MPs; however, bioaccumulation was predicted to decrease with increasing MPs ingestion. While ingestion of MPs has been demonstrated in nature, few studies have been conducted in freshwater environments and the adverse effects on organisms in their natural environment have not been well quantified.

One the main reason for the limited number of studies of MPs in natural ecosystems may be the low mass of particles that are collected during sampling. The analysis of POPs and other toxic compounds requires at least 1 g of MPs (or less if collected in an area with high concentration of toxic contaminants). While the reported number of MPs in some freshwater samples may range from 100–1000 particles, the total mass is generally less than 1 g due to the very small size of the MPs. The cost of chemical analysis and the sophisticated analytical instrumentation required also limits the number of analyses. Even though there have been more studies of MPs in marine systems than in freshwater, there are still only a limited number of reports of toxic compounds adsorbed onto MPs. The lack of harmonization of methodologies for collection and analysis, and no standard definition of sizes or categories of plastic particles are additional limitations for comparative analyses.

2.4 CONCLUSIONS

The number of studies on hazardous materials associated with MPs in both marine and freshwater has increased dramatically in recent years but additional data is still needed to fully understand the role that MPs play in transferring these materials from the environment to living organisms. Collection of basic information on the abundance and distribution of MPs in aquatic ecosystems is important; however, these collections should be combined with more advanced analyses to determine the concentrations of POPs, metals, bacteria and viruses adsorbed onto freshwater MPs. The effects of different types of synthetic plastic polymers, particle size and shape, and particle weathering on the rates of microbial and chemical adsorption/desorption should be determined in natural environments and not just in laboratory settings. Studies designed to investigate the rates of transfer of harmful materials from contaminated MPs to water and to aquatic organisms and the potential adverse effects of these compounds on organisms should be conducted under similar conditions to those seen in natural environments. More detailed information on the association of toxic compounds with freshwater MPs will help elucidate the real ecotoxicological threat of these materials to the aquatic environment and ultimately to human health.

REFERENCES

Ashton K., Holmes L. and Turner A. (2010). Association of metals with plastic production pellets in the marine environment. *Marine Pollution Bulletin*, **60**, 2050–2055.

Bakir A., Rowland S. J. and Thompson R. C. (2014). Enhanced desorption of persistent organic pollutants from microplastics under simulated physiological conditions. *Environmental Pollution*, **185**, 16–23. doi: 10.1016/j.envpol.2013.10.007.

Browne M. A., Galloway T. S. and Thompson R. C. (2010). Spatial Patterns of Plastic Debris along Estuarine Shorelines. *Environmental Science and Technology*, **44**(9), 3404–3409. DOI: 10.1021/es903784e.

Burns E. E. and Boxall A. B. A. (2018). Microplastics in the aquatic environment: evidence for or against adverse impacts and major knowledge gaps. *Environmental Toxicology and Chemistry*, **37**(11), 2776–2796. doi: 10.1002/etc.4268.

Chae Y. and An Y. (2017). Effects of micro- and nanoplastics on aquatic ecosystems: current research trends and perspectives. *Marine Pollution Bulletin*, **124**, 624–632.

Cole M., Lindeque P., Halsband C. and Galloway T. S. (2011). Microplastics as contaminants in the marine environment: a review. *Marine Pollution Bulletin*, **62**(12), 2588–2597. doi: 10.1016/j.marpolbul.2011.09.025.

Dang H. and Lovell C. R. (2000). Bacterial primary colonization and early succession on surfaces in marine waters as determined by amplified rRNA gene restriction analysis and sequence analysis of 16S rRNA genes. *Applied and Environmental Microbiology*, **66**(2), 467–475. doi: 10.1128/AEM.66.2.467-475.2000.

De Tender C. A., Devriese L. I., Haegeman A., Maes S., Ruttink T. and Dawyndt P. (2015). Bacterial community profiling of plastic litter in the Belgian part of the North Sea. *Environmental Science & Technology*, **49**(16), 9629–9638. doi: 10.1021/acs.est.5b01093.

do Sul J. A. I. and Costa M. F. (2014). The present and future of microplastic pollution in the marine environment. *Environmental Pollution*, **185**, 352–364. doi: 10.1016/j.envpol.2013.10.036.

Dobaradaran S., Schmidt T. C., Nabipour I., Khajeahmadi N., Tajbakhsh S., Saeedi R. and Faraji Ghasemi F. (2018). Characterization of plastic debris and association of metals with microplastics in coastline sediment along the Persian Gulf. *Waste Management*, **78**, 649–658. doi: 10.1016/j.wasman.2018.06.037.

Driedger A. G. J., Dur H. H., Mitchell K. and Van Cappellen P. (2015). Plastic debris in the Laurentian great lakes: a review. *Journal of Great Lakes Research*, **41**, 9–19.

Eerkes-Medrano D. and Thompson R. (2018). Occurrence, fate, and effect of microplastics in freshwater systems. Microplastic contamination in aquatic environments. In: Microplastic Contamination in Aquatic Environments: An Emerging Matter of Environmental Urgency, E. Y. Zeng (ed.), Elsevier, Amsterdam, Netherlands, pp. 95–132.

Eerkes-Medrano D., Thompson R. C. and Aldridge D. C. (2015). Microplastics in freshwater systems: a review of the emerging threats, identification of knowledge gaps and prioritization of research needs. *Water Research*, **75**, 63–82. doi: 10.1016/j.watres.2015.02.012.

Faure F., Demars C., Wieser O., Kunz M. and de Alencastro L. F. (2015). Plastic pollution in Swiss surface waters: nature and concentrations, interaction with pollutants. *Environmental Chemistry*, **12**(5), 582. doi: 10.1071/EN14218.

GESAMP (2015). Chapter 3.1.2 Defining 'microplastics'. Sources, Fate and Effects of Microplastics in the Mmarine environment: A Global Assessment. Chapter 3.1.2 Defining 'microplastics'. In: (IMO/FAO/UNESCO–IOC/UNIDO/WMO/IAEA/UN/UNEP/UNDP Joint Group of Experts on the Scientific Aspects of Marine Environmental Protection (GESAMP)), P. J. Kershaw (ed.), London, Rep. Stud. GESAMP No. 90, 96p, London.

Hahladakis J. N., Velis C. A., Weber R., Iacovidou E. and Purnell P. (2018). An overview of chemical additives present in plastics: migration, release, fate and environmental impact during their use, disposal and recycling. *Journal of Hazardous Materials*, **344**, 179–199. doi: 10.1016/j.jhazmat.2017.10.014.

Harmon S. M. (2018). The effects of microplastic pollution on aquatic organisms. In: Microplastic Contamination in Aquatic Environments: An Emerging Matter of Environmental Urgency, E. Y. Zeng (ed.), Elsevier, Amsterdam, Netherlands, pp. 249–270.

Hirai H., Takada H., Ogata Y., Yamashita R., Mizuhawa K., Saha M., Kwan C., Moore C., Gray H., Laursen D., Zettler E. R., Farrington J. W., Reddy C. M., Peacock E. E. and Ward M. W. (2011). Organic micropollutants in marine plastic debris from the open ocean and remote and urban beaches. *Marine Pollution Bulletin*, **62**, 1683–1692.

Holmes L. A., Turner A. and Thompson R. C. (2012). Adsorption of trace metals to plastic resin pellets in the marine environment. *Environmental Pollution*, **160**(1), 42–48. doi: 10.1016/j.envpol.2011.08.052.

Hong S. E., Shim W. J. and Jang M. (2018). Chemicals associated with marine plastic debris and microplastics: Analyses and contaminant levels. In: Microplastic Contamination in Aquatic Environments: An Emerging Matter of Urgency, E. Y. Zeng (ed.), Elsevier, Amsterdam, pp. 271–315.

Kirstein I. V., Kirmizi S., Wichels A., Garin-Fernandez A., Erler R., Löder M. and Gerdts G. (2016). Dangerous hitchhikers? evidence for potentially pathogenic vibrio spp. on microplastic particles. *Marine Environmental Research*, **120**, 1–8. doi: 10.1016/j.marenvres.2016.07.004.

Lagana P., Caruso G., Corsi I., Bergami E., Venuti V., Majolino D., La Ferla R., Azzaro M. and Cappello S. (2018). Do plastics serve as possible vector for the spread of antibiotic resistance? First insights from bacteria associated to a polystyrene piece from King George Island (Antarctica). *International Journal of Hygiene and Environmental Health*, **222**, 89–100. https://doi.org/10.1016/j.ijheh.2018.08.009.

Lee H., Lee H. and Kwon J. (2019). Estimating microplastic-bound intake of hydrophobic organic chemicals by fish using measured desorption rates to artificial gut fluid. *Science of the Total Environment*, **651**(Pt 1), 162–170. doi: 10.1016/j.scitotenv.2018.09.068

Li J., Zhang K. and Zhang H. (2018). Adsorption of antibiotics on microplastics. *Environmental Pollution*, **237**, 460–467.

Liebmann B., Köppel S., Königshofer P., Bucsics T., Reiberger T. and Schwabl P. (2018). Assessment of microplastic concentrations in human stool: Preliminary results of a prospective study. *United European Gastroenterology Journal*, **6**(Supplement 1), pp. A127.

Lim S., Yan B., Pitiranggon M., Li Y., McKee K., Gomes H. D. R. and Goes J. I. (2018). Distribution of Microplastics in the Estuarine Waters around the New York Metropolitan Area and Assessment of their Role as Potential Vectors of Toxic Compounds See: https://www.riverkeeper.org/wp-content/uploads/2018/04/Microplastics-Poster.pdf (accessed 2 May 2018).

Llorca M., Schirinzi G., Martínez M., Barcel D. and Farr M. (2018). Adsorption of perfluoroalkyl substances on microplastics under environmental conditions. *Environmental Pollution*, **235**, 680–691.

Lobelle D. and Cunlife M. (2011). Early microbial biofilm formation on marine plastic debris. *Marine pollution Bulletin*, **62**(1), 197–200.

Ma Y., Huang A., Cao S., Sun F., Wang L., Guo H. and Ji R. (2016). Effects of nanoplastics and microplastics on toxicity, bioaccumulation, and environmental fate of phenanthrene in fresh water. *Environmental Pollution*, **219**, 166–173.

Mato Y., Isobe T., Takada H., Kanehiro H., Ohtake C. and Kaminuma T. (2001). Plastic resin pellets as a transport medium for toxic chemicals in the marine environment. *Environmental Science and Technology*, **35**(2), 318–324. doi: 10.1021/es0010498.

McCormick A. R., Hoellein T. J., London M. G., Hittie J., Scott J. W. and Kelly J. J. (2016). Microplastic in surface waters of urban rivers: concentration, sources, and associated bacterial assemblages. *Ecosphere*, **7**(11), e01556. doi: 10.1002/ecs2.1556.

McCormick A., Hoellein T. J., Mason S. A., Schluep J. and Kelly J. J. (2014). Microplastic is an abundant and distinct microbial habitat in an urban river. *Environmental Science and Technology*, **48**(20), 11863–11871.

Millward G. E. (1995). Processes affecting trace element speciation in estuaries. A review. *Analyst*, **12**(3), 69–614. doi: 10.1039/AN9952000609.

Mincer T. J., Zettler E. R. and Amaral-Zettler L. A. (2016). Biofilms on plastic Debris and their influence on marine nutrient cycling, productivity, and hazardous chemical mobility. In: Hazardous Chemicals Associated with Plastics in the Marine Environment. The Handbook of Environmental Chemistry, H. Takada and H. Karapanagioti (eds.), Springer, Cham, vol. **78**.

Moore C. J., Lattin G. L. and Zellers A. E. (2011). Quantity and type of plastic debris flowing from two urban rivers to coastal waters and beaches of southern California. *Journal of Integrated Coastal Zone Management*, **11**(1), 65–73.

Prata J. C. (2018). Microplastics in wastewater: state of the knowledge on sources, fate and solutions. *Marine Pollution Bulletin*, **129**(1), 262–265. doi: 10.1016/j.marpolbul.2018.02.046.

Rech S., Macaya-Caquilpán V., Pantoja J. F., Rivadeneira M. M., Jofre Madariaga D. and Thiel M. (2014). Rivers as a source of marine litter – A study from the SE pacific. *Marine Pollution Bulletin*, **82**(1–2), 66–75. doi: 10.1016/j.marpolbul.2014.03.019.

Reisser J., Shaw J., Hallegraeff G., Proietti M., Barnes D. K.A. and Thums M. (2014). Millimeter-sized marine plastics: a new pelagic habitat for microorganisms and invertebrates. *PLoS ONE*, **9**(6), e100289. https://doi.org/10.1371/journal.pone.0100289

Rios L. M., Jones P. R., Moore C. and Narayan U. V. (2010). Quantitation of persistent organic pollutants adsorbed on plastic debris from the northern Pacific gyre's "eastern garbage patch" *Journal of Environmental Monitoring*, **12**(12), 2226–2236. doi: 10.1039/c0em00239a.

Rios L. M., Moore C. and Jones P. R. (2007). Persistent organic pollutants carried by synthetic polymers in the ocean environment. *Marine Pollution Bulletin*, **54**(8), 1230–1237. doi: 10.1016/j.marpolbul.2007.03.022.

Rios Mendoza L. M. and Evans C. Y. (2013). Plastics are invading not only the ocean but also the Great Lakes. 245th American Chemical Society National Meeting. April 7–11. New Orleans, LA.

Rios Mendoza L. M., Abebe F., Duhaime M. B. and Cable R. (2016). Microplastics as a source of Persistent Organic Pollutants in the Laurentian Great Lakes. IAGLR 59th Annual Conference on Great Lakes Research. http://iaglr.org/conference/downloads/2016_abstracts.pdf (accessed 6 June 2019).

Rios Mendoza L., Johnson C. and Nyeck Nyeck M. (2018). Macro and Microplastics: St. Louis River Estuary and Lake Superior. 6th International Marine Debris Conference. March 12–16. San Diego, CA. p 332 http://internationalmarinedebrisconference.org/wp-content/uploads/2018/10/6IMDC_Book-of-Abstracts_2018.pdf (accessed 8 June 2019).

Rochman C. M., Hoh E., Kurobe T. and Teh S. J. (2013). Ingested plastic transfers hazardous chemicals to fish and induces hepatic stress. *Scientific Reports*, **3**, 3263.

Teuten E. L., Rowland S. J., Galloway T. S. and Thompson R. C. (2007). Potential for plastics to transport hydrophobic contaminants. *Environmental Science & Technology*, **41**(22), 7759–7764. doi: 10.1021/es071737s.

UNEP (2016). Marine Plastic Debris and Microplastics – Global Lessons and Research to Inspire Action and Guide Policy Change. United Nations Environment Programme, Nairobi.

Van A., Rochman C. M., Flores E. M., Hill K. L., Vargas E., Vargas S. A. and Hoh E. (2012). Persistent organic pollutants in plastic marine debris found on beaches in San Diego, California. *Chemosphere*, **86**(3), 258–263. doi: 10.1016/j.chemosphere.2011.09.039.

Viršek M. K., Lovšin M. N., Koren Š., Kržan A. and Peterlin M. (2017). Microplastics as a vector for the transport of the bacterial fish pathogen species Aeromonas salmonicida. *Marine Pollution Bulletin*, **125**(1–2), 301–309. doi: 10.1016/j.marpolbul.2017.08.024.

Wang F., Wang F. and Zeng E. Y. (2018). Soprtion of toxic chemicals on micoplatistics. In: Microplastic Contamination in Aquatic Environments: An Emerging Matter of Environmental Urgency, E. Y. Zeng (ed.), Elsevier, Amsterdam, Netherlands, pp. 225–248.

Xie Z., Lu G., Yan Z., Liu J., Wang P. and Wang Y. (2017). Bioaccumulation and trophic transfer of pharmaceuticals in food webs from a large freshwater lake. *Environmental Pollution*, **222**, 356–366. doi: 10.1016/j.envpol.2016.12.026.

Zenker A., Cicero M. R., Prestinaci F., Bottoni P. and Carere M. (2014). Bioaccumulation and biomagnification potential of pharmaceuticals with a focus to the aquatic environment. *Journal of Environmental Management*, **133**, 378–387. doi: 10.1016/j.jenvman.2013.12.017.

Zettler E. R., Mincer T. J. and Amaral-Zettler L. A. (2013). Life in the "plastisphere": microbial communities on plastic marine debris. *Environmental Science & Technology*, **47**(13), 7137. doi: 10.1021/es401288x.

Chapter 3

Microplastics in wastewater treatment plants: A literature review of sampling methods and results

N. Mourgkogiannis and H. K. Karapanagioti

University of Patras, Department of Chemistry, Patras, Greece

Keywords: Average microplastic items per L, Continuous flow, Monitoring, Pharmaceutical microplastics, Sampling sites, Treatment stage sampling

3.1 INTRODUCTION

Plastic pollution in aquatic environments is not a new phenomenon but has become a global issue during the last decade. Municipal Wastewater Treatment Plants (WWTPs) should be considered as conduits for micro- and macroplastics (Mourgkogiannis *et al.*, 2018). Microplastics (MPs) cannot be easily detected by the naked eye, unlike macroplastics, and this fact makes their monitoring difficult. Moreover, analytical techniques, such as microscope and Fourier Transform Infrared Spectroscopy are required to detect their presence in wastewater.

MPs have been defined as plastic particles in a size range from 1 nm to 5 mm and several studies have shown that MPs can be found in oceans, on shores, in surface waters and in sediments around the world (GESAMP, 2015). Organic pollutants are sorbed by plastic particles as they float on the surface of polluted seas (Karapanagioti & Klontza, 2008; Ogata *et al.*, 2009). Moreover, plastics can interact with more than 600 marine organisms including fish, birds, mammals and others (Andrady & Rajapakse, 2017; Rochman *et al.*, 2013).

Over recent years, only limited studies have been able to quantify the presence of MPs and their release from WWTPs into receiving water systems. Samples used for

Editors: Hrissi K. Karapanagioti and Ioannis K. Kalavrouziotis
doi: 10.2166/9781789061697_0027

MPs monitoring have usually been taken from similar sampling sites such as pre-treatment or final effluent etc., but the sampling equipment used was different (electric pumps, stainless steel baskets, glass jars, etc.). Furthermore, samples were collected in a certain time period (e.g. dry period) with a certain frequency in order to avoid altering the volume of the collected effluent samples. It is also important that sampling equipment is rinsed, usually with distilled water, several times before collecting samples.

The aim of this chapter is to review the existing literature on the presence of MPs in WWTPs and their potential release to receiving waters. The specific objectives are (a) to present and evaluate the common techniques for collecting samples of wastewater effluent and (b) to determine the types and quantities of MPs that are released to receiving water bodies.

3.2 MICROPLASTICS IN WWTPs

The number of WWTPs increases globally due to population growth. Wastewater treatment should be considered necessary to treat influent wastewaters arising from households, sewers and industries by removing organic loads and pathogenic microorganisms to save and protect receiving water bodies such as rivers, lakes and seas which are used for drinking water, fishing, and for other water activities. The main question is whether WWTPs can also remove MPs from influents and stop their further entry into surface waters. For this reason, several studies have been reviewed. It can be stated that most of the facilities studied were not able to deal with MPs and remove them from wastewater.

MPs end up in WWTPs either from solid wastes from toilet systems or in sewer systems transferring wastewater and stormwater together via a combined system (Karapanagioti, 2017). Multiple studies around the world (extending from the USA to Korea, and from Greece to Finland) have shown and confirmed the presence of MPs in WWTPs and their escape into receiving waters (Gatidou *et al.*, 2018). Both primary and secondary MPs are released on a daily basis to aquatic environments and the quantities of escaped particles depends not only on the treatment stages of the WWTPs but also on the size of the WWTP filters, as well as on the size and range of MPs (Mourgkogiannis *et al.*, 2018). Browne *et al.* (2011) pointed out the presence of MPs in facilities and moreover calculated the quantities of released particles such as synthetic fibres in an advanced WWTP in Australia. Ziajahromi *et al.* (2017) observed that tertiary treatment facilities release smaller numbers of MPs than secondary treatment plants into receiving water bodies, on a daily basis. On the other hand, Murphy *et al.* (2016) pointed out the fate of MPs in one WWTP and concluded that despite the high removal efficiency of MPs many particles were released to the aquatic environment because of the large daily volume of effluent.

3.2.1 Sampling sites and methods for the detection of microplastics in WWTPs

For a researcher to collect samples of wastewater, the treatment stage to be sampled (stage sampling) and the sampling methods for analysing the samples for MPs are important. Over the years, sampling methods in WWTPs around the world have differed and, therefore, the outcomes from several studies are difficult to compare.

In early publications, researchers took samples for MPs from several different stages of treatment. Most commonly, samples were taken from pre-treatment and the chlorination tank to better estimate the quantities of MPs entering into and being released from the WWTP to receiving waters. Samples were also taken from other treatment stages such as from the aeration tank, secondary sedimentation tank and from sludge, to quantify MPs. Today, samples in most studies are collected from several treatment stages (pre-treatment, primary sedimentation tank, secondary sedimentation tank, etc.). However, the sampling method is not the same for every stage, and can be either continuous (direct) or random (indirect).

The direct sample collection method (continuous sampling) refers to samples being collected more than once with a certain frequency. Collecting tools such as electric pumps and faucets can be used according to this method. During the indirect collection method (random sampling), samples are collected just once in a random time using tools such as stainless steel and glass jars. Over the last few years, direct methods have been used more than indirect ones. The main difference between these two collection methods is that, with the direct method, researchers can confirm their results with repetitions compared to random sampling where results are given for a significant time period, disregarding parameters such as mobile equivalent population and seasons (during the summer season individuals move to coastal areas and so the WWTPs of these areas have to serve higher populations).

Although collecting instruments vary from one researcher to another, pumps are commonly used in many studies (Dris *et al.*, 2015; Mason *et al.*, 2016; Mintenig *et al.*, 2017; Talvitie *et al.*, 2015; Talvitie *et al.*, 2017a, b; etc.). In some other studies, such as Magnusson and Noren (2014), a Ruttner sampler was used. A Ruttner sampler is a water sampler consisting of a cylinder which is lowered into the water and closed with a plummet and is suitable for taking samples in lakes, WWTPs, etc.; the water sample volume range is 1.0–5.0 L and the height of the sampler varies from 56 to 92 cm, depending on the model.

As summarized in Table 3.1, most researchers refer either to random or sequential sampling, and a few like Talvitie *et al.* (2017a, b) used both sampling methods. Talvitie *et al.* (2017a) sampled in a time range of seven days from influent, after pre-treatment, after the activated sludge (AS) process, effluent, excess sludge, reject water, dried sludge via three different methods;

Table 3.1 List of representative results for microplastics (MPs) and sampling methods used in WWTPs.

Reference	Country/Area	Number of WWTPs	Level of Treatment in Municipal WWTP	Equivalent Population	Sample Collecting Instruments	Effluent Flowrate	Sampling Sites	Sampling Methods	
								Random	Continuous
Gies *et al.* (2018)	Canada, Vancouver	1	Secondary	1.3×10^6	Teledyne ISCO glacier portable water sampler (15 L buckets)	$\approx 493 \times 10^3$ m^3/d	• Primary effluent • Secondary effluent • Primary sludge • Secondary sludge		• 1 Lfor influent • 30 L for primary and secondary • 1 L wastewater collected into glass jars • 30 L passed through a 63 μm stainless steel sieve
Gundogdu *et al.* (2018)	Turkey	2	Secondary	1.5×10^6	Endress + Hauser ASP-Station 2000	178 ± 6 m^3/d	• After rough filter • Final effluent	5 L/d	
Lares *et al.* (2018)	Finland Mikkeli	1	• Primary • Secondary	55,000	Stainless steel buckets (10 L)	10,000 m^3/d	• Influent (after screening) • Effluent after primary clarification • Effluent after disinfection • Activated sludge (after aeration basin) • MBR sludge • Digested sludge	(4.0–30.0 L) (once in every two weeks for 3 months)	
Simon *et al.* (2018)	Denmark	10	• 9 Municipal • 1 Industrial (75%)	1.5×10^6	Glass bottles	$\approx 1.15 \times 10^6$ m^3/d	• Inlet • Outlet	Raw wastewater:1 L Effluent: 4.1–81.5 L	
Dyachenko *et al.* (2017)	USA, East Bay	1	Secondary	680,000	Faucet	227,000 m^3/d	After S. effluent		2 h/450 L 24 h/5,400 L
Leslie *et al.* (2017)	The Netherlands, Rhine and Meuse rivers in the Netherlands and Germany	7	• Westpoort (R1) • Amsterdam West facility (R2) • Amstelveen (R3) • Blaricum (R4) • Horstermeer (R5) • Houtrust (R6) • Heenvliet (R7)	NA	Glass jars	NA	• R1, R2: effluents and sewage sludge were sampled. • R3, R4, R5, R7: samples of both influents and effluents were taken. • R6: effluent only was collected. • R7: experimenting with a membrane (MBR)	2 L	
Mintenig *et al.* (2017)	Germany, Lower Saxony	12	• Secondary • Tertiary	NA	Mobile pumping device	5×10^{-4} 36×10^3 m^3/d	• Either at the overflow of clarifying tanks or at the intake of maturation ponds • Before and after post filtration (only at the largest WWTP)	1 m^3 (stopped in case of a reduction of flow rate)	

Talvitie *et al.* (2017a)	Finland, Viikinmaki (Helsinki region)	1	Tertiary	800,000	Electric pump (Biltema art.17–953)		270,000 m^3/d	• Influent • After pretreatment • After the AS process • Plant effluent • Excess sludge • Reject water • Dried sludge	Grab Sampling: NA quantity	**Composite** Sampling (volume with 20 mm filter): • Influent 0.1 L • After pre-treatment 0.2 L • After AS 1 L • Effluent 2 L **Sequential sampling** (volume with 20 mm filter): • Influent 0.1 L • After pre-treatment 0.2 L • After AS 1 L • Effluent 2 L
Talvitie *et al.* (2017b)	Finland	4	Tertiary	NA	Electric pump (Biltema art.17-953) Faucet	RSF, DAF DF, MBR	NA	Micro-screen filtration with discfilters with pore size 10 μm (DF10) Micro-screen filtration with discfilters with pore size 20 μm (DF20) Rapidy Sand Filters (RSF) Dissolved Air Flotation (DAF) Membrane bioreactor (MBR)	Sample volumes after the filters DF 10 300 μm 6-50 L 100 μm 6-50 L 20 μm 2 L DF20 300 μm 50 L 100 μm 50 L 20 μm 2 L RSF 300 μm 1000 L 100 μm 1000 L 20 μm 70 L DAF 300 μm 1000 L 100 μm 300 L 20 μm 2 L MBR 300 μm 140 L 100 μm 140 L 20 μm 140 L	24 h composite sample RSF 300 μm 25.5 L 100 μm 25.5 L 20 μm 25.5 L DAF 300 μm 16.1 L 100 μm 16.1 L 20 μm 16.1 L MBR 300 μm 6.1 L 100 μm 6.1 L 20 μm 6.1 L
Ziajahromi *et al.* (2017	Australia	3	• Primary • Secondary • Tertiary	≈1.5×10^6	Via gravity		• Primary: 308×10^3 m^3/d • Secondary: 17×10^3 m^3/d • Tertiary: 13×10^3 m^3/d	Primary: Post primary Secondary: Post primary and Secondary treatment Tertiary: Post primary, Tertiary and RO treatment		• Primary: 3 L in 25 μm • Secondary: 27 L in 25 μm • Tertiary: 200 L in 25 μm
Carr *et al.* (2016)	USA, Southern California	8	• 7 Tertiary • 1 Secondary	≈6.1×10^7	NA		NA	• Influent pumps • Primary settling tank • Primary solids • Aeration tank • Activated sludge • Final settling tank • Gravity filters		1.89×10^5–2.32×10^5 L

(*Continued*)

Table 3.1 List of representative results for microplastics (MPs) and sampling methods used in WWTPs. (*Continued*)

Reference	Country/Area	Number of WWTPs	Level of Treatment in Municipal WWTP	Equivalent Population	Sample Collecting Instruments	Effluent Flowrate	Sampling Sites	Sampling Methods	
								Random	Continuous
Houtz *et al.* (2016)	USA, San Francisco Bay	8	• Secondary • Tertiary	NA	Polypropylene containers	1.1 m^3/d	Final effluent	0.5–1 L	
Mason *et al.* (2016)	USA	17	NA	6.1×10^7	Extraction pump	$\approx 2 \times 10^6\ m^3/d$	Effluent		$5 \times 10^2 - 2.10 \times 10^4$ L
Michielssen *et al.* (2016)	USA, Detroit and Northfield	3	• Secondary (Detroit) • Tertiary (Northfield)	3×10^6	NA	• Detroit WWTP: $\approx 3 \times 10^9\ m^3/day$ • Northfield WWTP: $\approx 8 \times 10^6\ m^3/day$	• Raw wastewater 1–2 L • Preliminary effluent 1–6 L • Primary effluent 10–20 L • Secondary effluent 10–20 L • Final effluent: 34–38 L	NA	
Murphy *et al.* (2016)	Scotland, Glasgow	1	Secondary	650,000	Steel buckets (10 L)	261,000 m^3/d	• Influent • Grit & grease • Primary effluent • Final effluent		• Influent :30 L • Grit &Grease:30 L • Primary Effluent.: 30 L • Final Effluent.: 50 L
Dris *et al.* (2015)	France, Paris	1	Secondary	800,000	Automatic sampler	240,000 m^3/d	• Influent • Primary effluent • Final effluent		0.05 L of each sample (from 24 h average samples)
Talvitie *et al.* (2015)	Finland, Viikinmaki (Helsinki region)	1	Tertiary	840,000	Electric pump (pump drive 5206 Heidolph) Flow rate 1 mL/min	270,000 m^3/d	• Influent • After primary sedimentation • After secondary sedimentation • Purified wastewater (three replicates of each sample)	1.0×10^{-3} L/min	
Magnusson and Noren (2014)	Sweden, Långeviksverket in Lysekil	1	• Mechanically • Chemically • Biologically	14,000	Ruttner sampler	$\approx 2{,}100\ m^3/d$	• Influent water • Effluent water		• Influent: 3 replicates of 2 L • Effluent: 4 replicates of 1 m^3
Browne *et al.* (2011)	Australia, West Hornsby, Hornsby Heights	2	Tertiary	NA	Glass bottles (750 mL)	• West Hornsby: $\approx 274\ m^3/d$ • Hornsby Heights: $\approx 658\ m^3/d$	Discharge	3.75 L	

NA: Not available; AS: Activate Sludge; MBR: Membrane Bioreactor; RO: Reverse Osmosis.

grab sampling (samples at a certain time), 24-h composite samples, and 24-h sequential samples. In grab sampling, three replicate samples were collected from wastewater at the sampling sites by an electric pump at a depth 1 m. The disadvantage of this method is that, due to the high concentration of organics, the filters become clogged and only small volumes of samples can be collected. In composite sampling, samples were collected from different spots by an automatic water sampler with sampling at 15 min intervals over a time period of 24 h. Plastic containers containing the collected samples were placed in refrigerators. In sequential sampling, the samplers took 24 samples at 1-h interval. After sampling, 3 samples (3 h) were pooled together, resulting in 8 samples per sampling.

Gies *et al.* (2018) sampled from a secondary WWTP which served a more than 1.3 million equivalent population close to Vancouver in Canada. The collected samples were from primary and secondary effluent and from primary and secondary sludge. Talvitie *et al.* (2017a) sampled in a tertiary WWTP near Viikinmak in Finland which served a more than 800,000 equivalent population. The samples were collected from influent, after pre-treatment, after the activated sludge (AS) process, from the effluent, excess sludge, reject water and dried sludge. The most significant observation to be made about these two studies is that both the sampling location and level of treatment varied amongst the WWTPs. Similarly, Michielssen *et al.* (2016) took samples from a secondary WWTP (Detroit) and from a tertiary WWTP (Northfield); the sample volumes differed between each sampling site, being 1 to 2 L from raw wastewater, 1 to 6 L from preliminary effluent, 10 to 20 L from primary effluent, 10 to 20 L from secondary effluent, and 34 to 38 L from the final effluent.

Researchers used random collecting (indirect) methods. For example, steel buckets (10 L) were used to collect samples from treatment sites without any sequence. Murphy *et al.* (2016) sampled from four points in a secondary WWTP located on the Clyde River, Glasgow, Scotland, sampling 30 L from the first three points with the use of 10 L steel buckets. One of these four points was between the coarse and fine screening, another after the grit and grease chamber, and another after the primary sedimentation tank. Finally, a 50-L sample was collected from the plant effluent.

Dyachenko *et al.* (2017) sampled MPs in a secondary WWTP in East Bay in the USA. Their sampling method was based on a flow-through system which provided a continuous pass of effluent through stainless steel sieves either for a time period of 24 h or a sequential pass of effluent for a particular time in a 24-h period. These two studies and their results were milestones in developing sampling methods for measuring MPs in wastewater. The continuous flow sampling of sewage (direct method) is an evolution in MPs monitoring in these facilities. Moreover, both studies concluded that a considerably large number of MPs enter the receiving water environment.

3.2.2 Type and quantities of MPs released from WWTPs

Primary and secondary MPs have been found in WWTPs, and some are routinely released to aquatic environments. MPs, such as synthetic fibres, microbeads, spherical particles and others can be found in facilities all over the world but the quantities of MPs released to receiving waters differ from country to country and from year to year. Carr *et al.* (2016) concluded that no MPs were found after tertiary treatment and that secondary treatment plants (1440 MPs/L) could be considered as a more probable source of MPs than tertiary ones in the USA after sampling in seven tertiary and one secondary WWTP. On the other hand, Talvitie *et al.* (2017a) in Finland sampled one tertiary WWTP where small quantities of MPs (0.7–3.5 MPs/L) were detected after tertiary treatment.

The presence of MPs in WWTPs is already known and has been studied by the scientific community, as seen in Tables 3.1 and 3.2. Globally, a moderate range of studies have been completed and show that the presence and type of MPs differs among WWTPs. To detect the type of MPs, scientists passed the sample volumes through either stainless-steel sieves or filters with different pore sizes. The most common stainless-steel sieve diameter was 8 cm (Carr *et al.*, 2016; Dyachenko *et al.*, 2017) whilst Ziajahromi *et al.* (2017) used sieves with a diameter of 12 cm. Pore size, as mentioned above, varies at different stages of treatment. Usually, according to the MP size range to be detected, the pore size of the sieves used varied from 5 mm (Dyachenko *et al.*, 2017; Lares *et al.*, 2018) to 0.7 μm (Leslie *et al.*, 2017).

In a recent study, Herzke *et al.* (2018) pointed out the presence of MPs and fibres contained in wastewater effluent along the coast of Svalbard (Norway). MPs in WWTPs should be considered as a major priority issue representing a real threat to aquatic systems. Ramírez-Álvarez *et al.* (2018) evaluated the presence and the impact of MPs in Todos Santos Bay in Mexico and concluded that WWTPs are the main source of plastic microfragments in the Bay (with 75×10^4 to 196×10^4 MPs per h).

Wastewater consists not only of water but also of organic and inorganic substances and microorganisms. These three components show high concentrations in the first stages of treatment processes and are reduced as the fluid moves through the treatment stages. They are considered to be responsible for the clogging of filters during sampling when the effluent passes through the sieves or filters. This phenomenon, the clogging of sieve pores, is quite common and referred to in many studies. Researchers consume time and effort while facing such clogging problems during their research. So far, no solution has been found to sieve a large volume of effluent without the sieves clogging, unlike with the analysis of samples where the organics are digested either through a chemical or an enzymatic method (Dyachenko *et al.*, 2017). This situation is a major concern for researchers because they are forced to sample minimum volumes of fluid most of the time. Particularly in the pre-treatment and primary stages of

Table 3.2 List of representative results formicroplastic (MPs) types and quantities in WWTPs.

Study Reference	Stainless – Mesh Sieves- Filter			Type of MPs	Results
	Number	Type of Filter	Pore Size		
Gies *et al.* (2018)	1	Steel sieves	63 μm	• MP particles • Fibers	0.5 MP/L
Lares *et al.* (2018)	2	Steel sieves	5 mm 0.25 mm	• Fibers • Fragments	1.05 MP/L
	1	Vacuum filtration cellulose filter glass fiber (at the bottom)	0.8 μm 1.5 μm		
Gundogdu *et al.* 2018	1	Steel sieves	55 μm	• Fibers (polyester) • Plastic fragments • Piece of film	1.8–7.8 MP/L
Simon *et al.* (2018)	1 (Ø 47 mm)	Stainless sieves	10 μm	• PS microbeads • HDPE particles	19–447 MP/L
Dyachenko *et al.* (2017)	4 (Ø 8 cm)	Stainless sieves	5 mm 1 mm 0.355 mm 0.125 mm	• Pellet fragments • Bead fragments	• Continuous flow for 5 monthly 24 h sampling: 0.34 MP/L • Continuous flow for a single 2 h sampling: 2.4 MP/L
		Buchner Funnel (vacuum filtration) (cellulose)	0.8 μm (Ø 90 mm)		
Leslie *et al.* (2017)	1	Filters (Glass fiber)	0.7 μm	• Fibers • Foils • Spheres	9–91 MP/L
Mintenig *et al.* (2017)	1	Stainless sieves	10 μm	Synthetic fibers (polyester)	0.1–10.5 MP/L

(*Continued*)

Table 3.2 List of representative results formicroplastic (MPs) types and quantities in WWTPs. (*Continued*)

Study Reference	Stainless – Mesh Sieves- Filter			Type of MPs	Results
	Number	Type of Filter	Pore Size		
Talvitie *et al.* (2017a)	1	Filter device	300 μm 100 μm 20 μm	• Polyethylene fragments • Cleansing scrubs	0.0005–0.3 MP/L
Talvitie *et al.* (2017b)	1	Filter device	300 μm 100 μm 20 μm	PES PE PP PS PU PVC PVA Polyamide Acrylamide Polyacrylates Alkyd resin PPO EVA	DF 10 μm 0.3 (±0.1) MP/L DF 20 μm 0.03 (±0.01) MP/L RSF 0.02 (±0.007) MP/L DAF 0.1 (±0.04) MP/L MBR 0.005 (±0.004) MP/L
Ziajahromi *et al.* (2017)	4(Ø 12 cm)	Stainless sieves	500 μm 190 μm 100 μm 25 μm	• Synthetic fibers • Microplastic beads	• Primary: 1.5 MP/L • Secondary: 0.48 MP/L • Tertiary: 0.28 MP/L
Carr *et al.* (2016)	4 (Ø 8 cm)	Stainless sieves	400 μm 180 μm 45 μm 20 μm	Blue polyethylene (toothpaste formulation)	0.001 MP/L
Houtz *et al.* (2016)	NA	NA	NA	Poly- and Perfluoroalkyl substances (PFASs)	0.071–0.19 MP/L

Mason *et al.* (2016)	2	Tyler sieves	0.355 mm 0.125 mm	• Microfibers • Microparticles	0.05 MP/L
Michielssen *et al.* (2016)	5	Stainless sieves	4.75 mm 0.85 mm 0.3 mm 0.106 mm 0.02 mm	• Fragments (rough, irregularly shaped) • Fibers (both single filaments and threads of multiple twisted filaments) • Paint chips • Microbeads (perfectly spherical)	0.5–5.9 MP/L
Murphy *et al.* (2016)	4	Stainless sieves Vacuum filtered Whatman No1 (cellulose)	65 μm 11μm (Ø 90 mm)	• Fibers • Bead • Flake	• Influent: 5.7 (±5.23) MP/L • Grit & grease: 8.7 (±1.56) MP/L • Primary effluent: 3.4 (±0.28) MP/L • Final effluent: 0.25 (±0.04) MP/L
Dris *et al.* (2015)	1	Whatman filters (Sigma–Aldrich) (glass fiber)	1.6 μm (Ø 13 mm)	• Fibers • Spherical particles	14–50 MP/L
Talvitie *et al.* (2015)	1	Filter device	200 μm 100 μm 20 μm	• Textile fibers • Synthetic particles	8.6 MP/L
Magnusson and Noren (2014)	1	Incoming water: Stainless steel fitted with plankton net	300 μm (Ø 80 mm)	• Fibers • Plastic fragments • Flakes	0.00825 MP/L
	1	Effluent: Stainless steel fitted with plankton net	300 μm (Ø 80 mm)		
Browne *et al.* (2011)	NA	NA	NA	Polyester (67%) Acrylic (17%) Fibers: polyamide (16%)	1 MP/L

NA: Not available; MP: Microplastics; PS: Polystyrene; HDPE: High Density Polyethylene; Ø: Diameter.

treatment, only a small volume of effluent can pass through sieves because of the high concentration of organic content. Murphy *et al.* (2016) were only able to filter 30 L of influent from the grit and grease chamber and from primary effluent compared to the 50 L they were able to filter from final effluent in a secondary WWTP in Scotland.

Primary and secondary MPs have been found in WWTPs and a measurable number is released daily to the aquatic environment. Fibres are the main and dominating microparticles above all others (Table 3.2). MPs, spherical particles, microbeads and cleansing scrubs are all among the fragments that have been mentioned in several studies and these particles can be found in many WWTPs around the world (Kalčíková *et al.* 2017). Their presence has been observed in all treatment stages but their number decreases from stage to stage until a small number of microparticles end up in the aquatic environment. Generally, the main observation is the decreasing occurrence of MPs released to receiving waters based on the increasing level of treatment of WWTPs. Dyachenko *et al.* (2017) found that, in a secondary WWTP, the concentration of fragments in effluent water was 1.4 MPs/L compared to Talvitie *et al.* (2017a) who, in a tertiary WWTP, found MP concentration in effluent to be 0.0005–0.3 MPs/L.

The size of detected MPs in influents and effluents depends on the pore size of the sieve. Both Dyachenko *et al.* (2017) and Murphy *et al.* (2016) used stainless steel sieves and vacuum filtration with a cellulose filter to filter samples. Dyachenko *et al.* (2017) used Whatman No. 1 (cellulose) filters with a pore size of 0.8 μm and counted an average of 1.4 MPs/L. Murphy *et al.* (2016) filtered the samples through Whatman No. 1 (cellulose) with a pore size of 11 μm and counted an average of 4.5 MPs/L. The main question is whether cellulose filters have any effect on the results of MPs presence in the effluent or not. Finally, according to the data given in Table 3.2, the average quantity of MPs found in effluents was calculated as 19.2 MPs/L in several studies.

3.3 MICROPLASTICS IN GREEK WWTPs

To our knowledge in Greece, only one study has been completed to research the presence of small plastic particles in WWTPs. Mourgkogiannis *et al.* (2018) studied a great number of WWTPs (101) in Greece for the existence of small plastic fragments visible with the naked eye, using questionnaires. Plastic particles, regardless of the type of sewage system (combined or separate) end up in treatment facilities. Particle dimensions and the screening size of pre-treatment affected the quantity of small plastic particles passing through the WWTPs and escaping to the water environment (Carr, 2017). Furthermore, the screen gap size in the majority of the WWTPs was not small enough to prevent the transfer of plastics to the aquatic environment. Moreover, small plastics, such as cotton swab sticks (a dominant pollutant), plastic caps, pellets, smaller parts of plastic bags, hair clumps and condoms can be detected in Greek

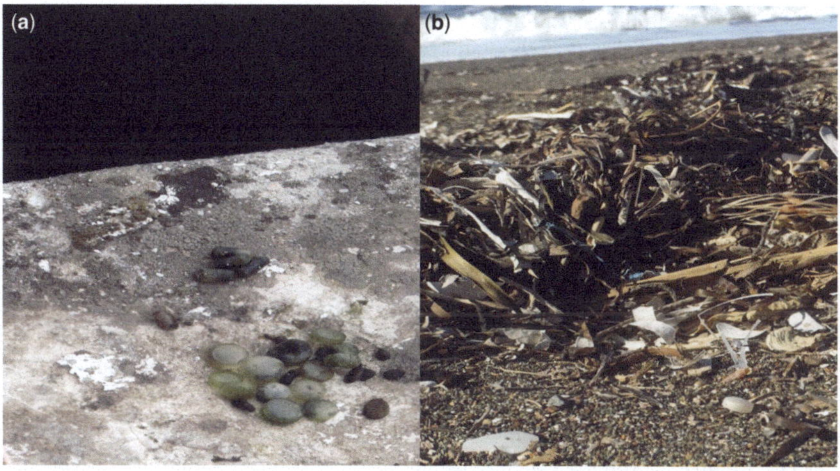

Figure 3.1 Pharmaceutical MPs found (a) in a WWTP and (b) on a beach close to the WWTP outlet.

WWTPs and on the sea shores next to the WWTP outlets. The presence of small plastics in WWTPs varies mainly according to population density, and to people's behaviours and habits.

Another type of particle observed in two WWTPs that were visited was pharmaceutical MPs (Figure 3.1 (a)). This is a form of MP that has so far not been studied extensively. They were found in the chlorination tank and the collecting bucket of a secondary sedimentary tank as well as on the beach close to the WWTP outlet (Figure 3.1 (b)). The sampling tools used were sieves with a pore size of 2 mm, and tweezers. These WWTPs provided secondary treatment and in different regions of Greece. The first was in Mainland Greece, whilst the other was in Western Greece and both served an equivalent population of 10,000. The sewer systems of these two WWTPs were connected to hospitals.

More research is still necessary because the presence, detection and quantification of MPs in Greek WWTPs are so far not well-studied. Studies will have great importance from a geographical point of view as Greece is surrounded by seas, and the outlets of the majority of Greek WWTPs end up in the sea, which could be directly contributing to marine pollution.

3.4 CONCLUSIONS

Even though there are several similarities between older and more recent studies conducted into the presence of MPs in WWTPs, the sampling methods, sampling locations and type of MPs which are detected in WWTPs vary considerably. Wastewater usually comprises large volumes of organic, inorganic and biological elements, especially in the initial stages of treatments in WWTPs. The condition of this wastewater is accepted as the main reason why small volumes of influents are filtered by collecting instruments during sampling. Most researchers use pumps (electric, mobile) with a continuous flow rate and a sampling frequency of defined time periods to collect samples. However, vacuum filtration with cellulose fibres is necessary to test cross-contamination of samples. The dominant MP fragments, identified in WWTPs, are synthetic fibres while others, such as microbeads, foils, spheres, microparticles and cleansing scrubs, can also be detected in final effluent. The average MPs released daily to the aquatic environment from 80 WWTPs in 19 studies worldwide was calculated as 19.2 MPs/L. The majority of the facilities were secondary treatment plants and just a few were tertiary systems. The removal efficiency of MPs in tertiary facilities is higher compared to secondary but, despite this high removal efficiency of WWTPs, many MPs are still released to water environments because of the high volume of wastewater daily entering each WWTP. Finally, common acceptable sampling methods, sampling instruments and sampling filtration should be adapted by the scientific community for future needs. To conclude, WWTPs should be considered as a source of primary and secondary MPs to the aquatic environment and a number of measures should be taken to prevent the existence of MPs in such receiving water bodies. So far, besides the above studies on MPs presence in WWTPs, no regulations have been announced to reduce the release of MPs from WWTPs into the environment. Consumer behaviour should be considered as an important factor and may have a significant role on MP reduction in aquatic systems (Karapanagioti & Kalavrouziotis, 2018).

ACKNOWLEDGMENTS

The authors would like to thank Prof. Ioannis K. Kalavrouziotis from the Hellenic Open University for his valuable contribution in writing this chapter.

REFERENCES

Andrady A. L. and Rajapakse N. (2017). Additives and chemicals in plastics. In: Hazardous Chemicals Associated with Plastics in the Marine Environment (Handbook of Environmental Chemistry 78), H. Takada and H. K. Karapanagioti (eds.), Springer International Publishing AG: pp. 1–18. DOI 10.1007/698_2016_124.

Browne M. A., Crump P., Niven S. J., Teuten E., Tonkin A., Galloway T. and Thompson R. (2011). Accumulation of microplastic on shorelines worldwide: sources and sinks. *Environmental Science & Technology*, **45**, 9175–9179.

Carr S. A. (2017). Sources and dispersive modes of micro-fibers in the environment. *Integrated Environmental Assessment and Management*, **13**(3), 466–469.

Carr S. A., Liu J. and Tesoro A. G. (2016). Transport and fate of microplastic particles in wastewater treatment plants. *Water Research*, **91**, 174–182.

Dris R., Gasperi J., Rocher V., Mohamed S., Renault N. and Tassin B. (2015). Microplastic contamination in an urban area: a case study in Greater Paris. *Environmental Chemistry*, **12**(5), 592–599.

Dyachenko A., Mitchell J. and Arsem N. (2017). Extraction and identification of microplastic particles from secondary wastewater treatment plant (WWTP) effluent. *Analytical Methods*, **9**, 1412–1418.

Gatidou G., Arvaniti S. O. and Stasinakis S. A. (2018). Review on the occurrence and fate of microplastics in sewage treatment plants. *Journal of Hazardous Materials*, **367**, 504–512.

GESAMP (2015). Chapter 3.1.2 Defining 'microplastics'. Sources, fate and effects of microplastics in the marine environment: a global assessment. In: (IMO/FAO/UNESCO–IOC/UNIDO/WMO/IAEA/UN/UNEP/UNDP Joint Group of Experts on the Scientific Aspects of Marine Environmental Protection (GESAMP)), P. J. Kershaw (ed.), Rep. Stud. GESAMP No. 90, 96p, London.

Gies E. A., LeNoble J. L., Noël M., Etemadifar A., Bishay F., Hall E. R. and Ross P. S. (2018). Retention of microplastics in a major secondary wastewater treatment plant in Vancouver, Canada. *Marine Pollution Bulletin*, **133**, 553–561.

Gundogdu S., Cevik C., Guzel E. and Kilercioglu S. (2018). Microplastics in municipal wastewater treatment plants in Turkey: a comparison of the influent and secondary effluent concentrations. *Environ Monit Assess*, **190**, 626.

Herzke D., Sundet J. H. and Jenssen M. (2018). Microplastics and fibres in the marine environment of Svalbard, Norway. In: Sixth International Marine Debris Conference (6IMDC) Book of Abstracts. 6IMDC, co-hosted by the National Oceanic and Atmospheric Administration (NOAA) and the United Nations Environment (UN Environment) in San Diego, California, 12–16 March 2018. p. 194. See: http://internationalmarinedebrisconference.org/wp-content/uploads/2018/10/6IMDC_Book-of-Abstracts_2018.pdf (accessed 6 June 2019).

Houtz E. F., Sutton R., Park J. S. and Sedlak M. (2016). Poly- and perfluoroalkyl substances in wastewater: Significance of unknown precursors, manufacturing shifts, and likely AFFF impacts. *Water Research*, **95**, 142–149.

Kalčíková G., Alič B., Skalar T., Bundschuh M. and Gotvajn A. Ž. (2017). Wastewater treatment plant effluents as source of cosmetic polyethylene microbeads to freshwater. *Chemosphere*, **188**, 25–31.

Karapanagioti H. K. (2017). Microplastics and synthetic fibers in treated wastewater and sludge. In: *Wastewater and Biosolids Management*, I. K. Kalavrouziotis (ed.), IWA Publishing, London, pp. 77–88.

Karapanagioti H. K. and Kalavrouziotis I. K. (2018). Microplastics in Wastewater Treatment Plants — A totally preventable source, In: Sixth International Marine Debris Conference (6IMDC) Book of Abstracts. 6IMDC, co-hosted by the National Oceanic and Atmospheric Administration (NOAA) and the United Nations Environment (UN Environment) in San Diego, California, 12–16 March 2018. p. 41. See: http://internationalmarinedebrisconference.org/wp-content/uploads/2018/10/6IMDC_Book-of-Abstracts_2018.pdf (accessed 6 June 2019).

Karapanagioti H. K. and Klontza I. (2008). Testing phenanthrene distribution properties of virgin plastic pellets and plastic eroded pellets found on Lesvos island beaches (Greece). *Marine Environmental Research*, **65**, 283–290.

Lares M., Ncibi M. C., Sillanpää M. and Sillanpää M. (2018). Occurrence, identification and removal of microplastic particles and fibers in conventional activated sludge process and advanced MBR technology. *Water Research*, **133**, 236–246.

Leslie H. A., Brandsma S. H. and Vethaak A. D. (2017). Microplastics en route: Field measurements in the Dutch river delta and Amsterdam canals, wastewater treatment plants, North Sea sediments and biota. *Environment International*, **101**, 133–142.

Magnusson K. and Noren F. (2014). Screening of microplastic particles in and down-stream a wastewater treatment plant. IVL-report C 55, Swedish Environmental Research Institute, Stockholm. See: https://www.diva-portal.org/smash/get/diva2:773505/FULLTEXT01.pdf (accessed 6 June 2019).

Mason S. A., Garneau D., Sutton R., Chu Y., Ehmann K., Barnes J., Fink P., Papazissimos D. and Rogers D. L. (2016). Microplastic pollution is a widely detected in US municipal wastewater treatment plant effluent. *Environmental Pollution*, **218**, 1045–1054.

Michielssen M. R., Michielssen E. R., Ni J. and Duhaime M. B. (2016). Fate of microplastics and other small anthropogenic litter (SAL) in wastewater treatment plants depends on unit processes employed. *Environmental Science Water Research & Technology*, **2**, 1064–1073.

Mintenig S. M., Int-Veen I., Löder M. G. J., Primpke S. and Gerdts G. (2017). Identification of microplastic in effluents of waste water treatment plants using focal plane array-based micro-Fourier-transform infrared imaging. *Water Research*, **08**, 365–372.

Mourgkogiannis N., Kalavrouziotis I. K. and Karapanagioti H. K. (2018). Questionnaire-based survey to managers of 101 wastewater treatment plants in Greece confirms their potential as plastic marine litter sources. *Marine Pollution Bulletin*, **133**, 822–827.

Murphy F., Ewins C., Carbonnier F. and Quinn B. (2016). Wastewater Treatment Works (WwTW) as a Source of Microplastics in the Aquatic Environment. *Environmental Science & Technology*, **50**(11), 5800–8.

Ogata Y., Takada H., Mizukawa K., Hirai H., Iwasa S., Endo S., Mato Y., Saha M., Okuda K., Nakashima A., Murakami M., Zurcher N., Booyatumanondo R., Zakaria M. P., Dung L. Q., Gordon M., Miguez C., Suzuki S., Moore C. J., Karapanagioti H. K., Weerts S., McClurg T., Burresm E., Smith W., Van Velkenburg M., Lang J. S., Lang R. C., Laursen D., Danner B., Stewardson N. and Thompson R. C. (2009). International Pellet Watch: global monitoring of persistent organic pollutants (POPs) in coastal waters. 1. Initial phase data on PCBs, DDTs, and HCHs. *Marine Pollution Bulletin*, **58**, 1437–1446.

Ramirez-Alvarez N., Rios-Mendoza L. M., Macías-Zamora J. V., Álvarez-Aguilar A., Oregel-Vázquez L., Hernández-Guzmán F. A., Sánchez-Osorio J. L., Charles M. J., Silva-Jiménez H. and Navarro-Olache L. F. (2018). Microplastic distribution in environmental matrices (water-sediment) in Todos Santos Bay, Mexico. In: Sixth International Marine Debris Conference (6IMDC) Book of Abstracts. 6IMDC, co-hosted by the National Oceanic and Atmospheric Administration (NOAA) and the United Nations Environment (UN Environment) in San Diego, California, 12–16 March 2018. p. 195. See: http://internationalmarinedebrisconference.org/wp-content/uploads/2018/10/6IMDC_Book-of-Abstracts_2018.pdf (accessed 6 June 2019).

Rochman C. M., Browne M. A., Halpern B. S., Hentschel B. T., Hoh E., Karapanagioti H. K., Rios-Mendoza L. M., Takada H., Teh S. and Thompson R. C. (2013). Classify plastic waste as hazardous. *Nature*, **494**, 169–171.

Simon M., van Alst N. and Vollertsen J. (2018). Quantification of microplastic mass and removal rates at wastewater treatment plants applying Focal Plane Array (FPA)-based Fourier Transform Infrared (FT-IR) imaging. *Water Research*, **142**, 1–9.

Talvitie J., Heinonen M., Pääkkönen J. P., Vahtera E., Mikola A., Setälä O. and Vahala R. (2015). Do wastewater treatment plants act as a potential point source of microplastics? Preliminary study in the coastal Gulf of Finland, Baltic Sea. *Water Science & Technology*, **2**(9), 1495–504.

Talvitie J., Mikola A., Setälä O., Heinonen M. and Koistinen A. (2017a). How well is microlitter purified from wastewater? A detailed study on the stepwise removal of microlitter in a tertiary level wastewater treatment plant. *Water Research*, **109**, 164–172.

Talvitie J., Mikola A., Koistinen A. and Setälä O. (2017b). Solutions to microplastic pollution – Removal of microplastics from wastewater effluent with advanced wastewater treatment technologies. *Water Research*, **123**, 401–407.

Ziajahromi S., Neale P. A., Rintoul L. and Leusch F. D. (2017). Wastewater treatment plants as a pathway for microplastics: development of a new approach to sample wastewater-based microplastics. *Water Research*, **112**, 93–99.

Chapter 4

Microplastics: Transport and removal at wastewater treatment plants

S. A. Carr[1] *and J. Thompson*[2]

[1]*Sanitation Districts of Los Angeles County, San Jose Creek Water Quality Control Laboratory, Whittier, California, USA*
[2]*Sanitation Districts of Los Angeles County, San Jose Creek Treatment Plant, Whittier, California, USA*

Keywords: Biofilm, Buoyancy, Density, Gravitational partitioning, Lipophilicity, Polymers, Skimming

4.1 INTRODUCTION

To date there have been few comprehensive studies of the impact various treatment stages at wastewater treatment plants have on the conveyance and removal of microplastics. This lack of information has limited our understanding of the role different stages in the process stream may have in the removal of plastic residues in wastewater treatment plants (WWTPs). This chapter examines and reviews the processes that contribute to generic removal of solids during wastewater treatment and assesses how effective these existing schemes are at removing microplastics (GESAMP, 2015) in effluent streams. These findings are based on effluent studies conducted at seven tertiary and one secondary plant in Southern California.

Polymeric residues make up only a minor fraction of the solids that are conveyed to WWTPs daily (Carr *et al.*, 2016; Horton & Dixon, 2018; Talvitie *et al.*, 2017a, b). Our focus will, therefore, be on commonly utilized mechanical, chemical and biological processes associated with conventional activated sludge (CAS) treatment at WWTPs and how these existing processes can sustain the microplastic removal levels seen in recently published studies. It may appear

Editors: Hrissi K. Karapanagioti and Ioannis K. Kalavrouziotis
doi: 10.2166/9781789061697_0045

counter intuitive that plants which exploit only common bulk physical properties can still be effective at removing "new classes" of emerging micropollutants. However, recent studies suggest that the same solids isolation processes, which have been used in wastewater treatment for over a century, may still be the most reliable and effective approaches for isolating and removing microplastics in waste streams (Murphy *et al.*, 2016; Carr *et al.*, 2016). It may appear too good to be true that such simple methods which make use of only fractional density differences in waste streams can facilitate the isolation and removal of emerging plastic pollutants.

Other processes such as flocculation, initiated through the addition of chemical coagulants, can also assist bulk gravitational partitioning of colloidal size and neutrally buoyant solid particles in effluents (Bagchia *et al.*, 2016; Leslie *et al.*, 2017). It may be a mistake to conclude that common solids separation schemes are ineffective at removing microplastic and microfiber residues because they do not appear to be "state of the art", "modern" or "advanced" (Ziajahromi *et al.*, 2016, 2017; Simon *et al.*, 2018). It is our hope that a closer examination of existing unit processes will challenge these assumptions (Baldwin *et al.*, 2016; Schneiderman, 2015).

WWTPs – also called sewage treatment plants, water pollution control plants or water recovery plants – remove the majority of pollutants from wastewater before it is released to local receiving waters (Miller *et al.*, 2017; Hollender *et al.*, 2009; Mrowiec, 2018). At these facilities, physical and biological methods, which closely mimic the natural processes in rivers, streams, lakes and wetlands, are used to purify water. Natural waste removal processes that take weeks in the environment can be completed in only seven hours, on average, at modern waste

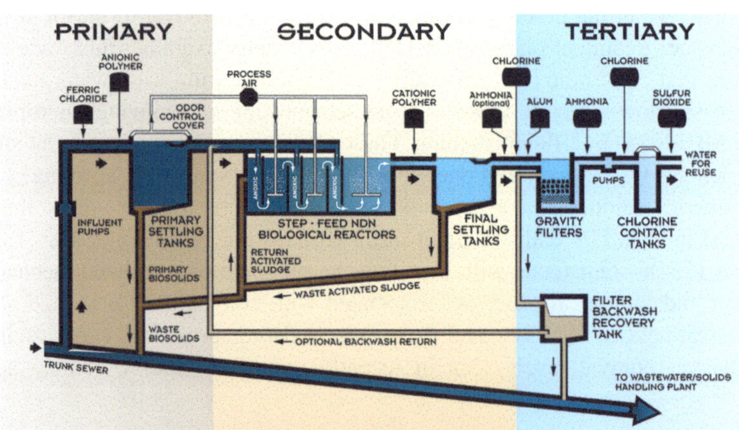

Figure 4.1 Tertiary WWTP schematic (without pretreatment or solid processing) taken from San Jose Creek plant flow diagram.
Source: County Sanitation Districts of Los Angeles County Public Information Office.

treatment facilities. These natural systems, by themselves, however, cannot handle the quantity of waste a modern metropolis produces (Clara *et al.*, 2005).

At WWTPs, the influent stream is subjected to six major unit processes: (i) preliminary treatment; (ii) primary treatment; (iii) secondary treatment; (iv) tertiary treatment (filtration); (v) disinfection; and (vi) sludge treatment (at solid handling facilities) (see Figure 4.1). Primary and secondary treatments remove about 85–95% of pollutants from the wastewater before it is disinfected and discharged into local waterways (Qasim, 1999).

Settled sludge, the byproduct of the primary treatment process, is digested for stabilization then dewatered to facilitate easier handling. This dewatered mass, known as biosolids, can then be applied to land as a soil amendment or processed further into compost or plant fertilizer.

4.2 PRELIMINARY TREATMENT

The treatment process is initiated when wastewater flows are conveyed to plants via sewer lines that are connected to homes and businesses. Plastic residues can, theoretically, be present in a wide array of sizes and types in waste flows, ranging in size from μm to cm or larger, just as in the natural environment. Incoming wastewater, referred to as "influent", initially passes through screens consisting of upright bars, spaced 2.5–10 cm apart. These grated barriers trap larger pieces of trash and debris, such as rags, sticks, newspaper, soft drink cans, bottles, plastic cups and other bulk items which can damage downstream equipment and interfere with the treatment processes. The initial physical screening isolates solids from the waste stream based solely on size. These processes require no design or special modifications to isolate plastic components. In the influent, any solid residue larger than 2.0 cm in diameter will be trapped on the bar screens and removed. Solids trapped on the screens are collected and disposed at a landfill. The main sewage pumps then lift the wastewater from the screening chambers to the surface level of the plant where primary treatment commences.

4.2.1 Primary treatment

Following coarse screening, the wastewater enters grit and primary settling chambers, also called sedimentation tanks, where it resides for one to two hours, depending on plant flows. At some facilities, pretreatment can be enhanced by the addition of alum or ferric chloride. The flow of the water at this stage is slowed to give heavier solids a chance to settle and the lighter materials the time to float. Both these partitioning processes occur simultaneously (Levine *et al.*, 1985) and, as presented below, these processes directly impact the removal of plastics. The floating mass (which is commonly referred to as "scum") is made up of grease, oils, plastics and soap. It is within this low density, hydrophobic aggregate that the full range of the common plastic residues that made it through the initial screening grates reside. This floating scum is removed by slow moving rakes or flights that continuously skim the surface of the tank.

4.2.2 Impact of the primary sedimentation stage on microplastics

4.2.2.1 High density plastics

The majority of plastics entering a WWTP are removed at the primary sedimentation stage. Plastics as well as other inorganic solid residues having densities >1.0 mg/L are isolated in the aerated grit chambers via settling. The higher density constituents, commonly referred to as "grit", is then pumped through cyclone degritters – devices that use centrifugal force to separate the sand, grit (such as coffee grounds), gravel and any high-density macro or microplastic residues. Applied aeration in the grit chambers improves settling of higher density solids, including plastics. The isolated grit is removed, washed and taken to landfills; the denser macro- and microplastics are disposed with this fraction. The partially treated wastewater then flows on to the Primary Sedimentation tanks for further separation and removal (see Tables 4.1 and 4.2).

Table 4.1 Microplastic estimates at selected plant stages.

Location	Sample Quantity	No. MPPs Counted	Estimated Total Daily MPP Counts at Solid Handling Plant
Primary Tank Skimming	5 g	20[a]	Total/Daily volumes not available
Secondary Tank Skimming	5 g	none found[a]	Total/Daily volumes not available
CST influent	100 mL	51	Total/Daily volumes not available
Thickened centrate	100 mL	267	Total/Daily volumes not available
Grit	2.1 g	1[a]	~7.78×10^6 based on 18 TPD
Biosolids	5 g	5[a]	~1.09×10^9 based on 1200 TPD
Final effluent	111,787 gallons	373	~0.93×10^6 based on 280 MGD
∑Grits + Biosolids + Final Effluent			1.099×10^9 per day
Grits + Biosolids			1.098×10^9 per day (~99.9% removal by the plant)
Influent			estimated 3.93 per gallon based on 280 MGD

[a]Average number found in 2 or 3 replicates.

Note: CST: centrate system treatment; TPD: tons per day; MGD: millions of gallons per day (1 million gallons = c.3785412 L); MPP: Microplastic particles.

Source: Carr *et al.* (2016).

Table 4.2 Plastic distribution in tertiary treatment plant.

Location	MPPs Count/Volume
Primary tank skimmings	Highest count[a]
Scum in aeration tanks	Some[a]
Return activated sludge	1 microplastic/20 mL[b]
Secondary effluent	1 microplastic/15,000 gallons
Gravity filter backwash	None found/12 gallons[b]
Final effluent	None found/50,898 gallons

[a]Could not be associated with an influent volume.
[b]Average of 4 replicates; MPP: microplastic particles; 1 Gallon = c. 3.79 L.

4.2.2.2 Low density plastics

Like omnipresent fats and oils from food waste, lower density microplastic separation is accomplished using only fractional density differences to effect partitioning in the primary tanks. A large fraction of plastics in common use is made up of low-density polyethylene and polypropylene. These plastics easily associate with hydrophobic residues such as fats, oils, grease and other lipophilic components and float to the top of primary settling tanks. The floating and separation steps do not appear to be impeded by the size of these hydrophobic components. Micro- and nano-oil droplets, whether dispersed as surfactant micelles or attached to other hydrophobic solids, fatty residues, congealed grease, microplastics and nanoplastics all appear to become associated within this amorphous aggregate. The entire floating mass is then removed by surface skimmers. Microscopic examination of this congealed mass appears to indicate that this facile process captures all hydrophobic residues, both liquids and solids, regardless of size.

Our initial plant studies confirmed that the vast majority of microplastics are removed at these primary treatment stages via skimming (see Figure 4.2 and Tables 4.1 and 4.2) and settling processes (Carr *et al.*, 2016). There was surprisingly little plastic observed in the secondary or latter treatment stages of the plant. These results mirrored the typical solid distribution and removal patterns in plants (Gies *et al.*, 2018).

4.3 SECONDARY TREATMENT

Because air and return activated sludge (RAS) are added to the wastewater to facilitate additional organic breakdown; secondary treatment is a type of activated sludge process. Air pumped into large aeration tanks adds oxygen, creates turbulence to mix the wastewater and sludge, and stimulates the growth of aerobic bacteria and other microbes in the sewage. These beneficial

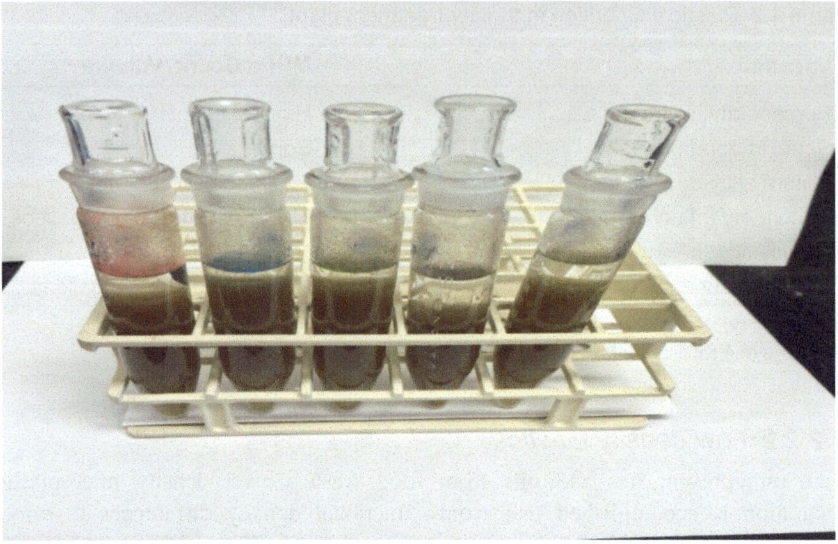

Figure 4.2 Various sized microplastics showing their partitioning and distribution in mixed liquor. Samples are from San Jose Creek Return Activated Sludge (RAS) mixed with micro polyethylene spheres: sizes (left to right): 10–45 μm (red); 53–63 μm (blue); 90–106 μm (green); 125–150 μm (violet); 250–300 μm (yellow).

microorganisms consume most of the remaining organic residues. Wastewater passes through the aeration tanks in three to six hours; this stage produces heavier particles that subsequently settle in the treatment process (Henze *et al.*, 2001). The aerated wastewater then flows to the final settling tanks which are similar to the primary settling tanks. Here the heavy particles and other dense solids settle to the bottom as secondary sludge. This sludge is re-circulated back to the aeration tanks to stimulate the activated sludge process. The returned sludge contains millions of microorganisms which sustain the right mix of bacteria in the aerated tanks to support the removal of a wide range of pollutants. In our studies the RAS contained surprisingly low plastic counts. The RAS solids accumulated over the mean cell residence time (MCRT) of 12 days from a 2.3×10^8 L/Day plant (total volume of 2.73×10^9 L). Because the total sludge volume at the plant studied was estimated to be $\sim 7.95 \times 10^7$ L, the theoretical solids concentration in the sludge should have been equivalent to ~34.3 times that of the average processed daily influent load. However, our RAS sample (a 20 mL aliquot) contained only one visible plastic fragment and not the 20.4 calculated, suggesting a removal efficiency >95% during primary treatment (see Table 4.2).

4.4 TERTIARY TREATMENT

4.4.1 The role of filters

One of the most interesting discoveries made in our study of tertiary plants was that filter beds used for final effluent polishing appeared to serve only a peripheral role in plastic removal in WWTPs. This observation was confirmed when large samples of filter backwash (45.43 L) were concentrated in the laboratory and then examined under a microscope. There was a surprising absence of microplastic residues in these backwash samples confirming that almost no microplastic residues were present or accumulating on filter bed surfaces. The suspicion that the filter beds were functioning as a final catch-all which limited microplastic discharges in the 227 MLD tertiary plant was not supported by these observations.

4.4.2 Disinfection using chlorine/hypochlorite

After primary and secondary treatment, pathogenic organisms may persist in the secondary effluent. Chlorine is typically added as a disinfectant to kill or inactivate any pathogens. Disinfection is thus an essential step to protect the public health of those using local beaches and enjoying other recreational activities at or near the wastewater discharge points. During disinfection, secondary effluent spends a minimum of 60–120 minutes in chlorine contact tanks in the presence of chloramines produced by the reaction of hypochlorite (the active component in common household bleach) with ammonia. The treated wastewater, or effluent, is then released to receiving waters. Common plastic containers are known to exhibit good resistance, even over long storage times, to concentrated disinfectants such as hypochlorite, chlorine dioxide and chloramines. Microplastics, by comparison, are exposed to highly diluted disinfectants in the low ppm range for relatively short times during wastewater disinfection. This thus strongly suggests that the prevailing attenuated plant exposures will be unlikely to facilitate transformation or effect structural changes in most common microplastics.

4.5 CHEMICAL AND MICROBIAL RESISTANCE OF POLYMERIC MATERIALS

4.5.1 Recalcitrant properties of plastics

Chemical attack on polymers can occur through hydrolysis, direct oxidation, photolysis etc. Highly reactive chemicals can also produce alterations on polymeric surfaces through modifications such as cross-linking, cyclization, breaking of polymeric chains, substitution and oxidative reactions (Burnett & Mark, 1954). After prolonged exposure, such events can lead to irreversible transformation of polymeric surfaces. However, when these reactions do occur

Table 4.3 Chemical resistance of plastics.

Substance class (at 20 °C)	LDPE	HDPE/XLPE	PP/PA	PMP	PEP/TFE/PFA	ECTFE/ETFE	ACL	PC	PSF	PVC	PS	PUR	NYL	PVDF	PMMA
Acids, dilute or weak	E	E	E	E	E	E	N	E	E	E	E	G	F	E	G
Acids, strong and concentrated	E	E	E	E	E	G	N	N	G	E	F	F	N	E	N
Alcohols, aliphatic	E	E	E	E	E	E	G	G	G	E	E	F	G	E	N
Aldehydes	G	G	G	G	E	E	F	F	F	N	N	G	F	E	G
Bases	E	E	E	E	E	E	G	N	E	E	E	N	F	E	F
Esters	G	G	G	G	E	E	G	N	N	N	N	N	E	G	N
Hydrocarbons, aliphatic	F	G	G	F	E	E	E	F	G	E	N	E	E	E	G
Hydrocarbons, aromatic	F	G	F	F	E	E	E	N	N	N	N	N	E	E	N
Hydrocarbons, halogenated	N	F	F	N	E	E	G	N	N	N	N	N	G	E	N
Ketones	G	G	G	F	E	G	G	N	N	N	N	N	E	N	N
Oxidizing agents, strong	F	F	F	F	E	F	N	N	G	G	N	N	N	G	N

Classification key:

E — 30 days of constant exposure cause no damage. Plastic may even tolerate it for years.

G — Little or no damage after 30 days of constant exposure to the reagent.

F — Some effect after 7 days of constant exposure to the reagent. Depending on the plastic, the effect may be crazing, cracking, loss of strength or discoloration. Solvents may cause softening, swelling and permeation losses with LDPE, HDPE, PP, PA and PMP. The solvent effects on these 5 plastics are usually reversible, and the polymers will usually return to their normal state after the solvent evaporates.

N — Not recommended for continuous use. Immediate damage may occur. Depending on the plastic, the effect will be a more severe crazing, cracking, loss of strength, discoloration, deformation, dissolution or permeation loss.

their rates are extremely slow compared to those of their monomers, and it should be noted that the reactivities of different polymers show little in common with those of their monomeric units (see Table 4.3). The persistence and ongoing accumulation of polymers in our environment can be directly linked to this intrinsic inertness.

The resistance of plastics to strong oxidants, UV photolysis, microbial attack and physical weathering can be attributed to the protected nature of polymeric bonds. Not surprisingly, the observed reactivity and appearance of plastics can also be influenced by variables such as chirality (R/S), conformational configurations (cis, trans) and the polymer's inherent glass transition temperature (Tg) (Table 4.4). Polypropylene and polyethylene, for example, are slowly attacked by oxidizing acids, such as nitric acid, and by non-oxidizing acids in the presence of oxidizers when carbonyl and sulfate groups are present. At a molecular level, the reactivity of any material is governed by its weakest bond; but in plastics there

Table 4.4 Physical properties of various plastics.

Polymer Acronym*	Max. Use Temp. °C	Brittleness Temp °C	Transparency	Specific Gravity	Flexibility	% Water Adsorption
LDPE	80	−100	Translucent	0.92	excel	<0.01
HDPE	120	−100	Translucent	0.95	rigid	<0.01
PP	135	0	Translucent	0.9	rigid	<0.02
PMP	175	20	Clear	0.83	rigid	<0.01
FEP	205	−270	Translucent	2.15	excel	<0.01
ETFE/ECTFE	150	−105	Translucent	1.7	mod	<0.1
PC	135	−135	Clear	1.2	rigid	0.35
PVC	70	−30	Clear	1.34	rigid	0.06
PA	121	−40	Translucent	0.9	mod	<.02
PSF	165	−100	Clear	1.24	rigid	0.3
ACL	121	270	Opaque	1.43	rigid	0.41
PFA	250	−270	Translucent	2.15	excel	<0.03
PUR	82	−70	Clear	1.2	excel	0.03
XLPE	100	−118	Translucent	0.93	rigid	<0.01
NYL	90	0	Translucent	1.13	rigid	1.3
PSF	90	100	Clear	1.05	rigid	0.05
PMMA	50	na	Clear	1.2	rigid	0.3
PVDF	110	−62	Translucent	1.75	excel	0.05

*Polymer acronyms: ACL acetal (polyoxymethylene); ECTFE Halar ECTFE (ethylene-chlorotrifluoroethylenecopolymer); ETFE Tefzel ETFE (ethylene-tetrafluoroethylene); FEP Teflon FEP (fluorinated ethylene propylene); HDPE high-density polyethylene; LDPE low-density polyethylene; NYL nylon (polyamide); PA polyallomer; PC polycarbonate; PFA Teflon PFA (perfluoro-alkoxy); PMMA polymethyl methacrylate; PMP polymethylpentene ("TPX"); PP polypropylene; PS polystyrene; PSF polypolysulfone; PUR polyurethane; PVC polyvinyl chloride; PVDF polyvinylidene fluoride; TFE Teflon TFE (tetrafluoroethylene); XLPE cross-linked high-density polyethylene.

are other factors that contribute to its inherent inertness. The woven, matted distribution of polymeric chains sterically shields the composite from chemical attack by limiting access to the structure's most reactive sites. This feature at a most elementary level can generally explain why polymers are always so much less reactive than their monomeric units. Other factors such as a polymer's degree of crystallinity, the level of intermolecular bonding and the abundance of unreactive, saturated covalent bonds uniquely combine to produce these refractory synergies.

To get further insight into the overwhelming complexity of plastics, we need only compare the properties of the crystalline and amorphous states of identical polymers. Polymers having identical chemical composition can have vastly different reactive and stability profiles. Semi-crystalline polymers typically display lower reactivity profiles than their amorphous counterparts. The regular symmetrical structure of polyamides (nylons), for example, exhibit greater molecular flexibility than their more crystalline forms. When polymeric forms transition to a state of higher crystallinity, its structure becomes more rigid. This allows the overlapping chains to form a more effective barrier to molecular diffusion of chemical agents such as solvents and gasses. In another example, polycarbonates are easily attacked by most common solvents due to their intermediate polarity which lowers the level of intermolecular attraction in the polymeric matrix. This limited flexibility and low intermolecular attraction combine to make polycarbonates rigid but lessens their ability to withstand attacks by surfactants and solvents.

Finally, a polymer's stability and reactivity profile can also be influenced by the presence of additives such as plasticizers, fillers, stabilizers and colorants (Campo, 2008). These may all introduce minor, mostly superficial changes in a polymer's reactive profile, especially towards biological attack or oxidative susceptibility during disinfection. Even when all these variables are cataloged it still may not be possible to provide definitive statements on the fate of most plastics in wastewater treatment plants. What can be conclusively stated, however, is that because most plastics present such formidable resistance to chemical and biological attack, their likely disintegration or transformation during the relatively short conveyance travel time through wastewater treatment plants will be insignificant or more likely nonexistent, given the relatively brief solids retention time (SRT) at these facilities.

4.5.2 Biological transformation of plastics during wastewater treatment

As concluded above, polymeric fragments, in general, will present daunting challenges to any microbes attempting to degrade or utilize them as a food source. The use of common endo- and exo-digestive processes such as phagocytosis and pinocytosis to facilitate digestion and utilization of plastics as food sources appear to be ineffective. When all the challenges discussed earlier are considered, it thus seems highly unlikely that microplastic residues can provide any post-digested net-energy positive gains, that might make them valued food sources. The overall thermodynamics appear unfavorable even for the degradation of partially oxygenated and non-halogenated polymers. Effective steric shielding will nullify even the most aggressive and potent biological enzymes. It is not surprising then, that aerobic and anaerobic digestion processes in WWTPs have little impact on the removal of micro- or macroplastic residues. In a recent study, Rom *et al.* (2017) examined the fate of polylactide (PLA) fibers that were exposed to activated sludge treatment under mesophilic (36°C) and thermophilic (56°C) conditions for up to 4 weeks. The results revealed that PLA was only minimally transformed in the activated sludge systems and confirmed that even when the prevailing biological conditions are aggressive, as is common under both mesophilic and thermophilic activated sludge systems, they are insufficient to promote biodegradation of PLA and other plastics.

4.5.3 Are there extant photolytic influences on plastics during wastewater treatment?

Plastics entering wastewater treatment facilities are likely to encounter only limited exposure to ultraviolet or visible light. There are few areas where exposure to direct sunlight is even possible. During the primary treatment stages, many of the tanks are covered to control odors and other ambient emissions. Plastics entering WWTPs will experience only minimal direct exposure to UV. Hydrophobic plastics residues in plants are also likely to be coated in fat, oil and grease residues. This, combined with the relatively short conveyance times, permits us to eliminate UV exposure as a factor which can contribute to the breakdown of plastics or affect their fate in treatment plants.

4.6 OTHER FACTORS INFLUENCING THE FATE OF PLASTICS IN WWTPs

4.6.1 Neutrally buoyant plastics

Although buoyancy may be overlooked as a factor in the conveyance of plastic residues in a plant, the apparent density of micro-residues could have a significant impact on microplastic removal. Most microplastic particles observed in the final discharge were observed to be surrounded by biofilm deposits. Minor changes in plastic densities, from contributions such as biofilm growth, mineral deposition and surfactant surface wetting, can significantly influence removal efficiencies. The buildup of biofilm on surfaces may alter the density of micro-particles. Similarly, the association of plastics with other hydrophobic waste components could also nominally impact separation dynamics during wastewater treatment. In such cases it appears that many discharged plastics evaded skimming and settling processes when their apparent densities were altered to a more neutrally buoyant range. In discharges where microplastics were most abundant, the plastic residues were estimated to have longer system residence times than upstream facilities where plastics were absent in the final discharges. Longer residence time would likely facilitate or promote surface growth or colonization (Harrison *et al.*, 2018).

4.6.2 Mechanical/chemical fracturing of plastic particles at treatment plants

There have been recent studies that suggest that the mechanical fracturing of microplastics is a possible contributor to changes in particle counts during passage through a treatment plant. However, the likelihood of achieving fracturing of particles that are already <5 mm in size seems unlikely. The dynamic collision of particles with walls and other physical obstructions in an effluent stream is unlikely to impart sufficient energies to result in mechanical fracturing of microplastics. There are some likely energetic impacts between microplastics and pump impellers that can possibly result in fragmentation but overall these events are not likely to change particulate counts to any measurable degree during the course of transmission through a plant. The relatively short residence time, the mild concentration of the disinfectants and the comparatively low temperatures in plants present little destructive or fracturing opportunities to degrade plastics during wastewater treatment.

4.7 SLUDGE TREATMENT

Sludge treatment stages ultimately determine the fate of microplastics at solid handling facilities. The following are typical stages of the sludge treatment process.

4.7.1 Thickening

The waste activated sludge produced by secondary treatment is approximately 99% water and must be concentrated to enable its further processing. Thickening tanks allow the sludge to collect, settle and separate from the water for up to 24 hours. The water is then sent back to the head of the plant, or to the aeration tanks for additional treatment. Any plastics in the waste activated sludge will be disproportionately partitioned to the settled phase where they will be separated and removed.

4.7.2 Digestion

After thickening, the primary sludge (raw settled and skimmings) and secondary (thickened waste activated sludge) is further treated to make it safer for the environment. The sludge is pumped to oxygen free tanks, called digesters, and heated to at least 95°F (35°C) for 15–20 days. This stimulates the growth of anaerobic bacteria, which consume organic material in the sludge. Unlike the bacteria in the aeration tanks, these bacteria thrive in an oxygen-free or "anaerobic" environment. The digestion process stabilizes the settled primary sludge and waste activated thickened sludge by converting much of the digestible material into water, carbon dioxide and methane gas. The black sludge that remains after digestion has the consistency of pea soup and has little odor. This is called digested sludge. Digested sludge is pumped from sludge storage tanks to a dewatering facility. The dewatering has the effect of concentrating microplastic residue s without transforming or degrading these pollutants.

4.7.3 Sludge dewatering

Dewatering reduces the liquid volume of sludge by about 90%. Digested sludge is sent through large centrifuges that operate like the spin cycle of a washing machine. The centrifugal force from the fast spinning of the centrifuges separates most of the water from the solids in the sludge, creating a substance known as biosolids. The water drawn from the spinning process is then returned to the head of the plant for reprocessing. Adding organic polymer improves the consistency of the "cake", resulting in a firmer, more manageable product. The biosolids cake is approximately 25–27% solid by weight. It is here that the majority of the microplastics removed from WWTPs will ultimately reside.

4.8 REMOVAL OF PLASTIC PARTICLES IN WWTPs

In general, treatment plants appear to make little distinction in the way their processes handle macro- and micro-solids. Treatment plant solids removal processes also makes no apparent distinction between natural and anthropogenic wastes or their sources. It is often stated that WWTPs were "not designed" to remove microplastics and microfibers in effluent streams. This, admittedly, is true and sounds like a reasonable assertion to support the conjectural sentiments of some who believe that plants, which employ only simple treatment schemes are incapable, if not wholly ineffective, at removing microplastics. The reality is that wastewater plants were never designed to target a specific type or class of solid waste. They are designed, however, to address and remove broad fractions of generic solids sharing common physical properties like density, hydrophobicity or other physical characteristics. Any property that differentiates solids from aqueous effluents can be utilized to effect practical separation. In reality, well operated, modern treatment plants do not need to incorporate special, or enhanced isolation processes such as micro or nano-filtration to effectively target the removal of plastics. Macro- or microplastic residues will all fit into one or more common solid categories (Vesilind, 2003). This was confirmed when small plastic residues (<0.5 mm) were followed through a WWTP (Friedler & Pisanty, 2006). Magnusson and Wahlberg (2014) also studied the fate of microplastic particles in the influent and effluent at three different Swedish WWTPs with mechanical, chemical and biological treatments; their study found that the removal of two different size fractions of particles, those between 5 mm > 300 μm and 5 mm > 20 μm, averaged 99.4% for all microplastic particles. Similarly, other studies conducted on 10 separate Danish WWTPs found the retention of microplastics by plants to be very high, in the range of 99.7% of the mass of the microplastic particles that enter the plants (Vollertsen & Hansen, 2017), which is similar to our own findings (Carr *et al.*, 2016).

4.9 CONCLUSIONS

WWTPs appear to be very capable of handling the complete range of solids entering their facilities; plastic particles are not an exception. As long as a solid possesses some inherent property that can be exploited to physically effect isolation from the waste stream it will either be filtered, become part of the settled sludge or float with the lower density scum. Once separated, these isolated phases become amenable to direct removal via simple physical and mechanical processes that efficiently eliminate floatable or settleable solids. Fortunately, none of these separation schemes seem to be impacted by particle size. A number of recent studies that looked at the transport of microplastic particles through wastewater treatment plants appear to confirm this. Removal of microplastics at these facilities has been shown to be very effective: 95–99% for all visible plastic sizes in the influent (Carr *et al.*, 2016; Magnusson & Norén, 2014; Magnusson *et al.*, 2016; Mason *et al.*, 2016; Murphy *et al.*, 2016). These numbers are remarkable for treatment processes that were "not specifically designed" to target microplastics.

These high removal efficiencies should give pause to growing demands for treatment plants to undergo immediate upgrades to reduce microplastic discharges. It should be noted that plastics make up only a minute fraction of solids in the influent stream and the existing solid removal processes, which make use of only basic physical processes, achieve removal efficiencies over 98% by utilizing only nominal density differences and gravity. It would therefore be difficult to justify significant expenditures to improve existing unit processes that already appear to remove over 98–99.9% of the plastics in influent streams.

Demands to incorporate "advanced filtration" and other "innovative treatment trains" into plant design may appear to be a bit gratuitous, considering these removal percentages. The operational challenges resulting from any such modifications are likely to be daunting, not to mention costly. The incorporation of filtration processes into functionally compliant plants will, at best, only provide marginal improvements in removal efficiencies and, at worst, create unintended operational disruptions. The latter will be especially noticeable if these modifications are sited at locations where the microplastic counts are highest. If such measures are adopted, plants will have to implement aggressive cleaning strategies to minimize flow restrictions and avoid other unforeseen operational impairments. These cleaning challenges and related maintenance issues would likely be costly and highly disruptive to routine plant functions. Based on our knowledge of existing filtration technologies, most of the proposed modifications could be subject to extreme surface fouling and have unintended long-term operational consequences. Until such issues are fully vetted and addressed, it is prudent to utilize caution. Existing unit processes still appear to be suitable and compliant with the widest array of micro- and macro-hydrophobic residues including oil droplets, surfactant micelles, fat residues and microplastic residues in wastewater effluents. The direct skimming of low-density hydrophobic hydrocarbons still appears to be the most effective approach for the isolation and removal of microplastic particles in wastewater treatment plants.

ACKNOWLEDGMENTS

The authors wish to thank the following for their assistance in reviewing and providing technical assistance in completing this chapter: Chris Wissman, Joe Gully, Stephen Johnson, Jeff Valdes, Martha Tremblay, Robert Ferrante, Michelle V. Carr and the County Sanitation Districts of Los Angeles County for supporting these efforts.

REFERENCES

Bagchia S., Probasco S., MardanDoost B. and Sturma B. S. (2016). Fate of microplastics in Water Resource Recovery Facilities (WRRFs) and National Environmental Loading Estimates. *Proceedings of the Water Environment Federation*, **2016**(7), 353–361.

Baldwin A. K., Corsi S. R. and Mason S. A. (2016). Plastic debris in 29 great lakes tributaries: relation to watershed attributes and hydrology. *Environ. Sci. Technol*, **50**(19), 10377–10385.

Burnett G. M. and Mark H. F. (1954). Mechanism of Polymer Reactions (Vol. 3). Interscience Publishers, New York.

Campo E. A. (2008). Selection of Polymeric Materials: How to Select Design Properties from Different Standards. William Andrew, Norwich NY, USA.

Carr S. A., Liu J. and Tesoro A. G. (2016). Transport and fate of microplastic particles in wastewater treatment plants. *Water Res*, **91**, 174–182.

Clara M., Kreuzinger N., Strenn B., Gans O. and Kroiss H. (2005). The solids retention time—a suitable design parameter to evaluate the capacity of wastewater treatment plants to remove micropollutants. *Water Research*, **39**(1), 97–106.

Friedler E. and Pisanty E. (2006). Effects of design flow and treatment level on construction and operation costs of municipal wastewater treatment plants and their implications on policy making. *Water Research*, **40**(20), 3751–3758.

GESAMP (2015). Chapter 3.1.2 Defining 'microplastics'. Sources, Fate and Effects of Microplastics in the Marine Environment: A Global Assessment. In: (IMO/FAO/UNESCO-IOC/UNIDO/WMO/IAEA/UN/UNEP/UNDP Joint Group of Experts on the Scientific Aspects of Marine Environmental Protection (GESAMP)), P. J. Kershaw (ed.), Rep. Stud. GESAMP No. 90, 96p, London.

Gies E. A., LeNoble J. L., Noël M., Etemadifar A., Bishay F., Hall E. R. and Ross P. S. (2018). Retention of microplastics in a major secondary wastewater treatment plant in Vancouver, Canada. *Marine Pollution Bulletin*, **133**, 553–561.

Harrison J. P., Hoellein T. J., Sapp M., Tagg A. S., Ju-Nam Y. and Ojeda J. J. (2018). Microplastic-associated biofilms: a comparison of freshwater and marine environments. In: Freshwater Microplastics. The Handbook of Environmental Chemistry. M. Wagner and S. Lambert (eds.), Springer, Cham, vol. **58**.

Henze M., Harremoes P., la Cour Jansen J. and Arvin E. (2001). Wastewater Treatment: Biological and Chemical Processes. Springer Science & Business Media, Berlin.

Hollender J., Zimmermann S. G., Koepke S., Krauss M., McArdell C. S., Ort C. and Siegrist H. (2009). Elimination of organic micropollutants in a municipal wastewater treatment plant upgraded with a full-scale post-ozonation followed by sand filtration. *Environmental science & technology*, **43**(20), 7862–7869.

Horton A. A. and Dixon S. J. (2018). Microplastics: An introduction to environmental transport processes. *Wiley Interdisciplinary Reviews: Water*, **5**(2), e1268.

Leslie H. A., Brandsma S. H., Van Velzen M. J. M. and Vethaak A. D. (2017). Microplastics en route: field measurements in the Dutch River Delta and Amsterdam Canals, wastewater treatment plants, North Sea Sediments and Biota. *Environment International*, **101**, 133–142.

Levine A. D., Tchobanoglous G. and Asano T. (1985). Characterization of the size distribution of contaminants in wastewater: treatment and reuse implications. *Journal (Water Pollution Control Federation)*, 805–816.

Magnusson K. and Norén F. (2014). Screening of microplastic particles in and down-stream a wastewater treatment plant. Swedish Environmental Protection Agency, Environmental Monitoring Unit SE-106 48 Stockholm, *Sweden, IVL Swedish Environmental Research Institute.*

Magnusson K. and Wahlberg C. (2014). Mikroskopiska skräppartiklar i vatten från avloppsreningsverk. *Rapport NR B*, **2208**, 33.

Magnusson K., Lloyd H. and Talvitie J. (2016). Microlitter in sewage treatment systems: A Nordic perspective on waste water treatment plants as pathways for microscopic anthropogenic particles to marine systems. Nordic Council of Ministers.

Mason S. A., Garneau D., Sutton R., Chu Y., Ehmann K., Barnes J. and Rogers D. L. (2016). Microplastic pollution is widely detected in US municipal wastewater treatment plant effluent. *Environmental Pollution*, **218**, 1045–1054.

Miller R. Z., Watts A. J., Winslow B. O., Galloway T. S. and Barrows A. P. (2017). Mountains to the sea: river study of plastic and non-plastic microfiber pollution in the northeast USA. *Marine pollution bulletin*, **124**(1), 245–251.

Mrowiec B. (2018). The role of wastewater treatment plants in surface water contamination by plastic pollutants. In E3S Web of Conferences, EDP Sciences, Bielsko-Biala, July 2018. Poland, Vol. **45**, p. 00054.

Murphy F., Ewins C., Carbonnier F. and Quinn B. (2016). Wastewater treatment works (WwTW) as a source of microplastics in the aquatic environment. *Environmental Science & Technology*, **50**(11), 5800–5808.

Nalgene Company/Sybron Corp (2019). Chemical Resistance of Plastics: Chemical compatibility chart. See: https://www.calpaclab.com/chemical-compatibility-charts/ (accessed May 2019).

Qasim S. R. (1999). Wastewater Treatment Plants: Planning. Design and Operation, **2**, CRC Press, London.

Rom M., Fabia J., Grübel K., Sarna E., Graczyk T. and Janicki J. (2017). Study of the biodegradability of polylactide fibers in wastewater treatment processes. *Polimery*, **62** (11–12), 834–840.

Schneiderman E. T. (2015). Discharging Microbeads to Our Waters: an Examination of Wastewater Treatment Plants in New York, 1–11. See: https://ag.ny.gov/pdfs/2015_Microbeads_Report_FINAL.pdf. (accessed May 2019).

Simon M., van Alst N. and Vollertsen J. (2018). Quantification of microplastic mass and removal rates at wastewater treatment plants applying Focal Plane Array (FPA)-based Fourier Transform Infrared (FT-IR) imaging. *Water Research*, **142**, 1–9.

Talvitie J., Mikola A., Setälä O., Heinonen M. and Koistinen A. (2017a). How well is microlitter purified from wastewater? A detailed study on the stepwise removal of microlitter in a tertiary level wastewater treatment plant. *Water research*, **109**, 164–172.

Talvitie J., Mikola A., Koistinen A. and Setälä O. (2017b). "Solutions to microplastic pollution–Removal of microplastics from wastewater effluent with advanced wastewater treatment technologies". *Water Research*, **123**, 401–407.

Vesilind P. (ed.) (2003). Wastewater Treatment Plant Design (Vol. 2). IWA Publishing, London.

Vollertsen J. and Hansen A. A. (2017). Microplastic in Danish wastewater: Sources, occurrences and fate press, Danish Environmental Protection Agency.

Ziajahromi S., Neale P. A. and Leusch F. D. (2016). Wastewater treatment plant effluent as a source of microplastics: review of the fate, chemical interactions and potential risks to aquatic organisms. *Water Science and Technology*, **74**(10), 2253–2269.

Ziajahromi S., Neale P. A., Rintoul L. and Leusch F. D. (2017). Wastewater treatment plants as a pathway for microplastics: development of a new approach to sample wastewater-based microplastics. *Water research*, **112**, 93–99.

Chapter 5

Method development for microplastic analysis in wastewater

A. Dyachenko[1]*, M. Lash*[1] *and N. Arsem*[1,2]

[1]*East Bay Municipal Utility District (EBMUD), Laboratory Services Division, 2020 Wake Ave., Oakland, CA, 94607, USA*
[2]*Bay Area Clean Water Agencies (BACWA) Microplastics Workgroup, USA*

Keywords: Effluent, Nile Red, Sampling, Spectroscopy, Wastewater treatment plant (WWTP)

5.1 INTRODUCTION

In recent years, multiple studies and reports have been published on microplastic particles in various wastewater treatment plant (WWTP) effluents around the world. However, currently, no standardized methods for sampling, sample preparation, identification and quantification of microplastics in wastewater have been released. This fact contributes to discrepancies across studies and also compromises data comparison. It has been established that methods developed for the determination of microplastics in surface waters and sediments are not suitable for a complex matrix such as secondary WWTP effluent (Dyachenko *et al.*, 2017). While it is expected that tertiary WWTP effluent is a substantially cleaner matrix, the majority of WWTP discharges are produced by secondary treatment.

Two important observations can be made on review of available data on microplastic particle counts in wastewater effluent (as summarized in Table 5.1). First, there is a trend of decreasing microplastic particle count with increased total

Editors: Hrissi K. Karapanagioti and Ioannis K. Kalavrouziotis
doi: 10.2166/9781789061697_0063

Table 5.1 List of recent published results of microplastic count in wastewater treatment plant final effluents with study design details.

Ref.	Location and Effluent Type	Sample Collection	Sample Preparation	Analytical Technique	QA/QC	Microparticle Count
Carr *et al.* (2016)	Los Angeles, CA, USA: Secondary-treated	Continuous flow (11.4–22.7 L/min)/ 423,000 L through stacked sieves (100–400 μm)	No additional preparation	Microscopy; FTIR	Laboratory Fortified Blank	0.00088/L
Dris *et al.* (2015)	Paris, France: Secondary-treated	24 h composite samples; analyzed 50 mL aliquots	Filter onto 1.6 μm glass fiber filter	Microscopy	Field Blank	14–50/L
Dyachenko *et al.* (2017)	Oakland, CA, USA: Secondary-treated	(1) Continuous flow (1 gal/min), 24 h/5,440 L (2) Continuous flow (1 gal/min), 2 h/450 L Through stacked sieves (125–1,000 μm)	Wet peroxide oxidation; filter onto 0.8 μm membrane filter	Microscopy; FTIR and Raman confirmation	Laboratory Fortified Blank	(1) 0.34/L (2) 2.4/L
Gies *et al.* (2018)	Vancouver, Canada: Secondary-treated	Grab/30 L Sieved (63 μm)	Wet peroxide oxidation; filter onto 1 μm PC membrane; density separation (oil extraction protocol)	Microscopy; FTIR confirmation	Air Blank; Matrix Spike	0.5/L

Lares *et al.* (2018)	Mikkeli, Finland: Secondary-treated and tertiary-treated (membrane bioreactor)	(1) Secondary: Grab/17.5–30 L (2) Tertiary: Grab/16–23.5 L Through 0.25–5 mm stacked sieves	Dried; wet peroxide oxidation; filter onto 0.8 µm cellulose nitrate filter	Microscopy; FTIR and Raman confirmation	Field Blank; Laboratory Fortified Blank	(1) 1.0/L (2) 0.4/L
Mason *et al.* (2016)	17 WWTPs across the USA: Secondary-treated and tertiary-treated (advanced filtration)	Continuous flow: 12–18 L/min for 2–24 h/500–41,000 L through stacked sieves (125–355 µm)	Wet peroxide oxidation; mounted in DI water in glass Petri dish	Microscopy	Field Blank	0.017–0.195/L
Mintenig *et al.* (2017)	12 WWTPs in Germany: Secondary-treated and tertiary-treated (filtration)	Continuous flow: 390–1,000 L through custom filter cartridge (20 µm)	Enzymatic digestion (protease, cellulase); Wet peroxide oxidation; $ZnCl_2$ density separation; 0.2 µm aluminium oxide filter	Microscopy; FPA micro-FTIR and ATR-FTIR confirmation	Laboratory Reagent Blank	0.01–9/L
Michielssen *et al.* (2016)	2 WWTPs in Michigan, USA: Secondary-treated and tertiary-treated (sand filtration)	Grab; 34–38 L	Stacked sieves (0.2–4.5 mm)	Microscopy	Laboratory Reagent Blank	0.59–37.4/L
Murphy *et al.* (2016)	Glasgow, Scotland: Secondary-treated	Grab; 50 L through 65 µm sieve	Filter onto 11 µm Whatman filter	Microscopy; FTIR confirmation	Air blank	0.25/L

(*Continued*)

Table 5.1 List of recent published results of microplastic count in wastewater treatment plant final effluents with study design details. (*Continued*)

Ref.	Location and Effluent Type	Sample Collection	Sample Preparation	Analytical Technique	QA/QC	Microparticle Count
Simon *et al.* (2018)	10 WWTPs in Denmark: Secondary-treated and tertiary-treated (sand filtration)	Continuous; 4.1–81.5 L	10 μm sieve; enzymatic digestion (cellulase); wet peroxide oxidation; wet-sieve (10–80 μm); suspended in ethanol	FPA-based FTIR	Matrix Spike; Method Blank	19–447/L
Sutton *et al.* (2016)	8 WWTPs, California USA: Secondary-treated and tertiary-treated (filtration)	Continuous flow: 314–1,250 L though a stacked sieve (125–355 μm)	Wet peroxide oxidation; stacked sieves (125–355 μm)	Microscopy	Laboratory Fortified Blank; Laboratory Reagent Blank	0.071–0.19/L
Talvitie *et al.* (2015)	Helsinki, Finland: Tertiary-treated (biological filtration)	1 mL/min continuous flow/575 L through custom stacked sieve device (20–300 μm)	No additional preparation	Microscopy	Field Blank	8.6/L
Ziajahromi *et al.* (2017)	3 WWTPs near Sydney, Australia: Secondary-treated and tertiary-treated (ultrafiltration, reverse osmosis and de-carbonation)	(1) Secondary: Continuous flow (10 L/min): 150 L (2) Tertiary: Continuous flow (10 L/min): 200 L Through stacked sieves (25–500 μm)	Wet peroxide oxidation; density separation; Rose Bengal stain	Microscopy; FTIR confirmation	Laboratory Fortified Blank; Laboratory Reagent Blank; Matrix Spike	(1) Secondary: 0.48/L (2) Tertiary: 0.28/L

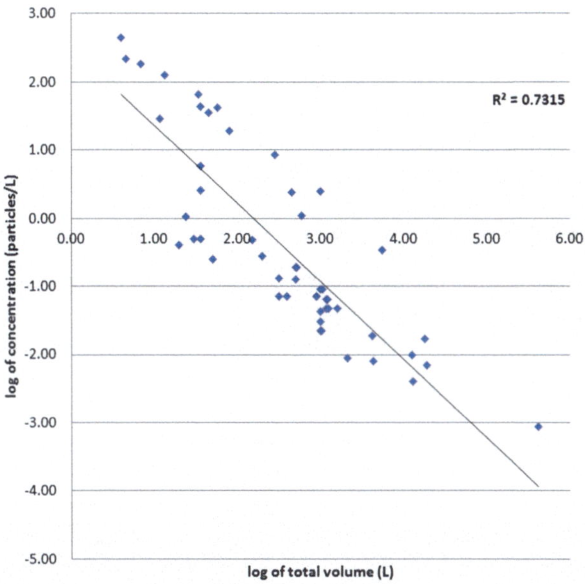

Figure 5.1 Logarithmic representation of total sample volume vs. calculated microplastic concentration, incorporating 51 secondary and tertiary final effluent samples pooled from 12 studies (Carr *et al.*, 2016; Dyachenko *et al.*, 2017; Gies *et al.*, 2018; Lares *et al.*, 2018; Mason *et al.*, 2016; Mintenig *et al.*, 2017; Murphy *et al.*, 2016; Simon *et al.*, 2018; Sutton *et al.*, 2016; Talvitie *et al.*, 2015, 2017; Ziajahromi *et al.*, 2017). Decreasing microplastic count trend as sampling volume increases with significant coefficient of correlation ($R^2 = 0.7315$).

effluent volume sampled, visualized in Figure 5.1. Grab samples and inadequate short sampling durations do not generate representative effluent samples that can be accurately used to extrapolate daily microplastic discharge and representative sampling is a key first step of method development.

Second, when only visual observation (microscopy) is used as an analytical tool, the number of microplastic particles tends to be higher when compared to spectroscopically confirmed results. It has been demonstrated that fibers smaller than 1.0 mm (such as cotton and nylon) can be visually indistinguishable (Dyachenko *et al.*, 2017). However, both negative and positive biases have been reported in studies that compared visual sorting of microplastics with spectroscopy (Hanvey *et al.*, 2016; Murphy *et al.*, 2016; Ziajahromi *et al.*, 2017).

Few, if any, scientific publications have comprehensively addressed the challenges of a standard method development for microplastic particles in wastewater. In particular, representative sampling considerations that take into account WWTP processes are rarely discussed. This chapter describes the challenges and steps necessary in the development of a standard method for quantitation of microplastics in secondary wastewater effluent. The following areas of microplastics identification in wastewater are considered significant factors that must be addressed during method development: sampling strategy (including sampling duration and total volume collected); sample preparation; analytical tools used for particle identification and enumeration; and quality control.

5.2 SAMPLING STRATEGY

As the initial step in microplastics analysis, collecting a representative wastewater sample is critical for accurate quantitation of microplastic particles and subsequent extrapolation to daily discharge. One important parameter that must be considered when attempting to collect a representative sample is WWTP hydraulic retention time (HRT), i.e. the amount of time it takes for wastewater to complete the treatment cycle and be discharged. Plant flow and retention times may vary greatly throughout the day based on many factors including consumer behavior, time of day, wet or dry weather conditions, time of year, WWTP capacity, maintenance operations, etc. The ability to collect a composite effluent sample continuously over a period of 24 h is one way to reduce sample variation and its dependence on WWTP retention times. Shorter duration sample collection techniques including discrete composite sampling (interval sampling) may lead to over- or underestimation of the number of microplastic particles detected, leading to error magnification when the results are extrapolated to daily discharge. Researchers have reported results for samples collected over a period of 1 or 2 h during peak personal product use hours; however, peak consumer use period or peak influent flow is often significantly different from peak effluent flow (Sutton *et al.*, 2016; Ziajahromi *et al.*, 2017). For example, Table 5.2 shows an actual measurement of daily WWTP activity during dry weather, assuming a 57 MGD plant flow resulting in total HRT of 12 h. Under these conditions, the wastewater stream entering the plant at 5:00 pm does not complete the treatment cycle until 5:00 am the following day.

Table 5.2 A secondary WWTP's hydraulic retention times (HRT).

Treatment Stage	Volume (MG)	No. in Service	Section Volume (MG)	HRT (h)
Sedimentation tanks	0.49	9	4.41	1.9
Reactors	1.58	5	7.90	3.3
Clarifiers	1.61	8	12.88	5.4
Effluent Pumping Station (EPS)	3.38	1	3.38	1.4
Total:			28.57	12

It is important to distinguish between a particular WWTP design and its actual operation at the time of sampling. More likely than not, WWTPs do not operate at full capacity at any given time. If it is not possible to perform continuous 24-h sampling, analysts should refer to historical values for best approximation of a representative sample. At a minimum, the diurnal extremes of plant flow and their timings must be considered when extrapolating results derived from shorter sampling periods. In addition to adjusting for temporal variation, standardizing sampling procedure is a key step in achieving consistent and reliable results. Secondary wastewater samples should be collected over a stack of sieves that are certified in mesh size and diameter. One of the major concerns when collecting secondary effluent is the quantity of interferences present, particularly cellulose, and fats, oil and grease (FOG). Furthermore, the amount of interference increases with sampling duration, with smaller mesh size sieves being especially prone to clogging as they accumulate most of the cellulose. Clogging during continuous flow collection may be avoided by stacking sieves of larger mesh sizes over the sample collection sieves, in descending order of mesh size, to screen out larger particles. For example, Dyachenko *et al.* (2017) stacked 5.0 and 1.0 mm sieves on top of two sample collection sieves (mesh size of 355 and 125 μm).

For sampling location, secondary effluent samples can be collected either before or after dechlorination depending on the plant's access point to the final effluent. Figure 5.2 shows a diagram of a WWTP with an effluent pumping station where samples can be continuously collected prior to dechlorination and discharge. The dechlorination stage is unlikely to contribute to microplastic concentration.

It is preferable to collect samples during dry weather to mitigate the impact from increased flows and the influence of storm water. In combined wastewater systems, storm water can carry street debris and affect the profile of the final effluent. A report should include weather conditions during sample collection. As a minimum, the effluent flow at the sampling point must be measured before and after sample collection. Ideally, a flow monitor should be employed to obtain an average reading throughout the duration of sampling. A simple way to measure the effluent flow is to fill a 10-L container and record the time taken to fill it. For accuracy, such a procedure should be repeated at least three times and the average value accepted (Sedlak *et al.*, 2017).

Recording accurate sample volume is essential since it is used in computation of the particle count discharged per day. Each facility must determine a maximum flow that does not result in clogging the larger sieve(s) to obtain the most representative sample; for example, at sites that typically have lower suspended solids in the final effluent, higher flow rates are appropriate for continuous sampling. An example of sampling documentation is shown in Figure 5.3. Once samples are collected by the sieves, the contents must be transferred to appropriate containers for sample processing. Using deionized water from a squirt bottle or dedicated hose, the analyst should transport all the material to one side of each sieve and wash it into

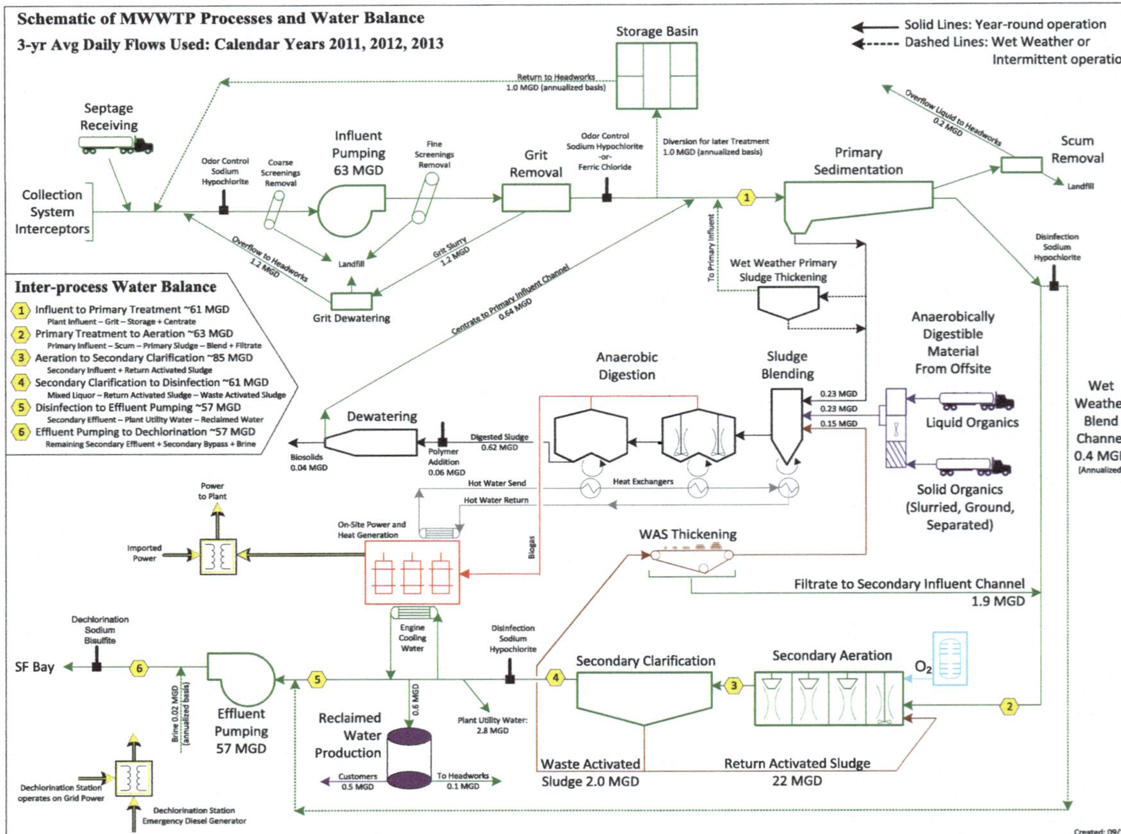

Figure 5.2 Schematic representation of East Bay Municipal Utility District's main WWTP's processes (Source: East Bay Municipal Utility District).

Wastewater Treatment Facility:		
Sample ID:	Start Date:	Start day of week:
Sample personnel/ WWTP Staff assisting:	Start time:	Stop time:
Field Observations:		
Sample port location (sink, exterior pipe etc.)	General Comments:	
Filenames of Any Photos Taken:		
Flow Measurement		
Method to determine flow (bucket/ time; meter, etc):	**Meter** Reading at start:	Reading at finish:
If using bucket, volume:	Start time:	Stop time:
If using bucket, volume:	Start time:	Stop time:
If using bucket, volume:	Start time:	Stop time:

Figure 5.3 WWTP microplastic sampling event field form (Sedlak *et al.*, 2017).

a glass sample jar of appropriate size. Additional tools such as tweezers or a spoon can be helpful in aiding the transfer with minimal loss of the collected material. All equipment must be thoroughly cleaned prior to use to avoid cross contamination. Sample containers should be labelled and stored in a temperature-controlled environment of 4–6°C to avoid potential bacterial growth. Hold times need to be established. Methanol or ethanol, which are not known to cause polymer degradation, can be used as a preservative to suppress bacterial growth during long-term storage and transport. Freezing samples should be avoided as it can lead to fracturing of microplastic particles.

5.3 SAMPLE PREPARATION

Wastewater samples are typically high in organic/biological matter that may interfere with the identification of microplastics. The amount of interference is proportional to the sample volume passed through the sieves. Secondary wastewater effluent has large amounts of cellulose and FOG interference that often bind to the surface of microparticles and need to be removed prior to analysis (Dyachenko *et al.*, 2017). Most common organic matter digestion techniques include wet peroxidation (WPO) including catalytic WPO (aka Fenton's reagent) and digestion with 10–20% potassium hydroxide (KOH) (Lares *et al.*, 2018; Silva *et al.*, 2018; etc.). Caution should be taken when subjecting samples to WPO at elevated temperatures as violent boiling may suddenly occur.

Enzymatic digestion has been shown to be effective and can be the first step in removing most persistent interferences. Cellulase enzyme has been shown to effectively digest cellulose whereas lipase enzyme should substantially reduce FOG. In addition, protease can effectively remove the remaining biological material. Density separation is not appropriate for secondary wastewater effluent since many organic and inorganic salts, coagulants and flocculants typically present in final effluent have densities similar to microplastics (Tagg *et al.*, 2015).

5.3.1 Chemical digestion

We have previously demonstrated insufficient digestion of cellulose using Fenton's reagent (Figure 5.4). Multiple digestion cycles may facilitate complete digestion of interferences but introduce the risk of particle loss, degradation and contamination. More recently, Munno *et al.* (2018) conducted a study of various digestion protocols for microplastics, which concluded that chemical digestion methods that apply heat greater than 60°C during the digestion or extract drying stages, will likely result in degradation of several polymer groups.

Digestion with 20% KOH at room temperature is effective in removing most organic matter but it may take up to seven or more days for complete digestion, especially for contents transferred from smaller sieves which accumulate most of the cellulose (Figure 5.5).

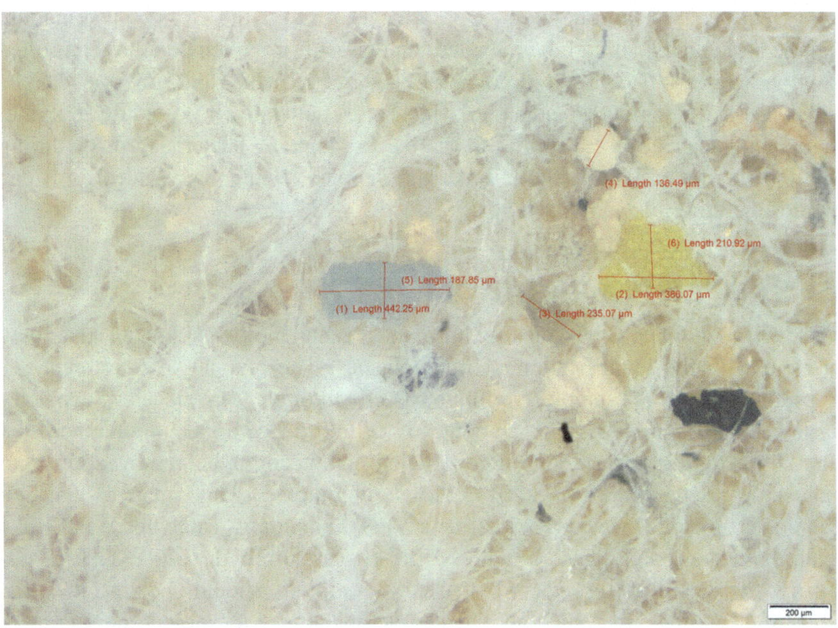

Figure 5.4 Secondary WWTP effluent extract, post WPO.

Figure 5.5 Secondary WWTP effluent before and after 7-day digestion with 20% KOH: (a) content from a 355 μm sieve; (b) content from a 125 μm sieve.

5.3.2 Enzymatic digestion

The most comprehensive enzyme-based protocol appears to be the one developed by Löder *et al.* (2017). The so-called Basic Enzymatic Purification Protocol (BEPP) includes sequential purification steps starting with sodium dodecyl sulfate (SDS) treatment followed by cellulase enzymatic digestion and H_2O_2 oxidation. The expanded Universal Enzymatic Purification Protocol (UEPP) appears to be especially appropriate due its use of additional enzymes such as protease and lipase. The latter are particularly effective in digesting FOG types of interference from a wastewater matrix. Due to potential competition and other interactions between various enzymes, the digestion must be sequential. On average, 24 h is allocated to each digestion step. This protocol has not been tested on composite secondary wastewater effluent samples.

One hybrid approach is to use enzymatic digestion to digest most of the cellulose and FOG followed by either 20% KOH or 30% H_2O_2 digestion. An inter-laboratory study using secondary effluent samples from various types of wastewater treatment needs to be conducted to determine the combination that is most effective for different types of discharges.

The final step of sample preparation is collecting the extracted particles onto a substrate which is typically achieved via vacuum filtration. The selected substrate must be compatible with the analytical technique e.g., aluminium oxide membrane filters are transparent to infrared waves and allow for robust Focal Plane Array (FPA) Micro-FTIR analysis in transmission mode (Loder *et al.*, 2017). Extracts should be stored in dark and dry conditions, free from cross-contamination prior to analysis.

5.4 PARTICLE ANALYSIS

Current analytical techniques employed for analysis of microplastics include microscopy, spectroscopy (FTIR, Raman), electron microscopy, chromatography (Pyrolysis-GC/MS and LC/MS) and fluorescent microscopy (Gago *et al.*, 2018; Li *et al.*, 2018; Zhang *et al.*, 2018).

Spectroscopic techniques such as Micro-FTIR and Micro-Raman are capable of producing accurate identification of microplastics by comparison of obtained spectra with reference library spectra or with the spectra obtained by analyzing in-house reference polymer standard materials. A typical particle size limitation of FTIR methodology is widely considered to be about 20–30 μm due to diffraction phenomenon. Higher resolution spectroscopy such as Raman is capable of identifying much smaller particles down to 1 μm and lower (Araujo *et al.*, 2018; Li *et al.*, 2018).

Although being the simplest method, visual observation (i.e. microparticle sorting using a stereomicroscope) is not sufficiently accurate as it is prone to misidentification error and over counting. These error types are accentuated by the complex wastewater matrix. Chromatographic techniques do not provide information about the number of microplastics or their morphology. Electron microscopy can be useful as a secondary analytical tool when spectroscopy does not provide a definite answer, particularly when inorganic salts are present.

5.4.1 Spectroscopic analysis

Spectroscopic analysis by Micro-FTIR is by far the more prevalent technique compared to Micro-Raman due to its lower cost and the availability of more developed standard reference spectral libraries. Raman cannot be used for fluorescent particles due to the competing nature of Raman scattering and fluorescence emission. For the purpose of a standard method development for microplastic particle identification in wastewater, a lower particle size quantitation limit of 100 μm is appropriate for spectroscopic analysis. While the typical instrument detection limit of Micro-FTIR is about 20–30 μm depending on configuration, it is a common practice for standard environmental water and wastewater methods to select a limit of quantitation at least 2–3 times higher than the instrument detection limit to account for matrix impact on sensitivity. Furthermore, visual evaluation and sorting with a microscope is often the first step preceding spectroscopic confirmation, with 100 μm being the practical lower limit for a typical 40x stereoscope. Until automated solutions are fully developed, it may not be feasible to confirm every particle spectroscopically. Particles should be categorized during visual observation by category: fiber, fragment, pellet, foam or film; and by color. Images of all particles should be recorded. One approach to streamline spectroscopic analysis is to confirm 10% of particles in any particular category if their number exceed 100, and to confirm 10 particles for any particular group of less than 100 particles.

A chemical spectrum is generated for each particle and compared to a known library of chemical spectra, and the method goal should be to pursue spectral matches that are 80% or greater. However, matrix interferences may make this challenging. Images should be recorded for each particle that is confirmed to be microplastic and its morphology documented. It has been reported that common chemical digestion techniques involving Fenton's reagent and 20% KOH as well as enzymatic digestion do not alter the infrared or Raman spectral properties of microplastic particles (Lares *et al.*, 2018; Mintenig *et al.*, 2017; Tagg *et al.*, 2015).

5.4.2 Automation

Micro-FTIR and Micro-Raman spectroscopy is a time-consuming analysis that requires highly trained analysts. Visual sorting and manual selection of particles for spectroscopic confirmation is a tedious and labor-intensive process. An automated approach which eliminates the visual sorting step was presented by Primpke *et al.* (2017) using Micro-FTIR combined with sophisticated software and the results obtained were in close agreement with manual analysis. However, such software is not yet commercially available.

5.4.3 Rapid screening/fluorescent microscopy

Visual screening of wastewater samples for microplastics is prone to misidentification due to the morphological similarities of some plastic and non-plastic particles, the potential transparency of polyethylene (PE) and other polymers, and interfering particles that physically mask target microplastics. The use of polymer-selective stains may enable more rapid and accurate screening. Nile Red is a lipophilic fluorescent stain whose emission spectrum changes with the polarity of its solvent and the stained particle. Nile Red can effectively stain microplastics (Andrady, 2011) and the resulting fluorescent signal can be used to detect potential microplastics in a complex matrix. Nile Red staining has been used effectively as a rapid screening tool for detection of microplastics in water and silt (Cole, 2016; Maes *et al.*, 2017; Shim *et al.*, 2016; Wang *et al.*, 2018) but differential fluorescent staining applied to wastewater samples has not yet been reported.

The methodology for use of Nile Red to screen microplastics in surface water and silt varies across studies. Stain concentration, carrier solvent, contact time, excitation wavelengths and microscope power all affect the effectiveness of this technique. The use of a stereoscope with a forensic light source to illuminate the sample by Maes *et al.* (2017) allowed for a large working distance and the ability to screen larger particles than by using a fluorescence microscope.

In studies using Nile Red to screen microplastics, the stain is most commonly dissolved in acetone, chloroform or methanol. However, acetone is known to dissolve polymers and Nile Red dissolved in acetone produced strong fluorescence with all microplastics tested but distorted smaller particles and melted polyacrylamide. Methanol produced weak fluorescence in target particles, and chloroform weaker still. Decreasing contact time with solvents decreased the distortion of microplastics but also resulted in a weaker fluorescent signal.

A procedure optimized by A. Dyachenko and M. Lash (unpublished data) for Nile Red staining is described as follows. A processed sample filtered onto a 37 mm glass fiber filter was placed in a glass Petri dish and treated with 1.2 mL of 5.0 μg/mL Nile Red in 1:1 methanol:acetone, then incubated with the lid on at 60°C for 30 minutes. This procedure resulted in strong fluorescence with negligible distortion of the target particles and minimal background staining of non-target particles. Using this procedure, six different plastics fluoresced bright yellow-orange when observed under a stereoscope, illuminated with a forensic light source of blue-green wavelengths (450–510 nm) and an orange (529 nm) filter (Figure 5.6). Background material was visible as a faint red-pink. Matrix spikes using 200–400 μm polystyrene beads were stained using this protocol and resulted in bright fluorescence of the beads, which were readily visible against the background. Transparent films became apparent following application of Nile Red (Figure 5.7). These films are suspected to be PE and fibers which are difficult to observe under white light illumination.

Samples intended for Nile Red staining should be prepared using glass fiber or aluminium oxide filters. Polycarbonate (PC), modified cellulose ester (MCE) and polyethersulfone (PES) filters distort when exposed to acetone or 1:1 acetone:methanol, and cellulose filters may retain stain. In secondary effluent, interferences following sample preparation may include cellulose, non-plastic fibers, organic and inorganic salt deposits, and fatty acids, depending on the digestion technique utilized. The staining procedure described above did not produce fluorescence in stearic acid, cellulose or cotton fibers in laboratory fortified blanks.

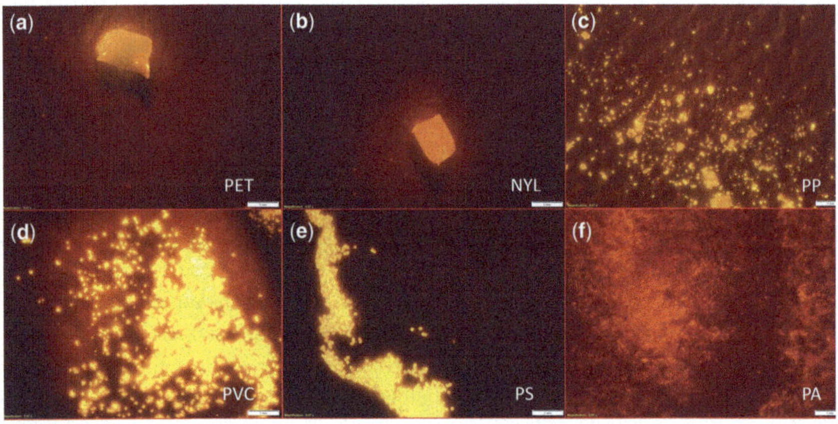

Figure 5.6 Laboratory fortified blanks (LFBs) stained with Nile Red and illuminated with a blue-green fluorescent light source viewed through an orange filter, magnified 6.7x. Top row (a–c): a polyethylene terephthalate pellet, a nylon pellet, polypropylene powder. Bottom row (d–f): polyvinyl chloride pellets, polystyrene beads, polyacrylamide powder.

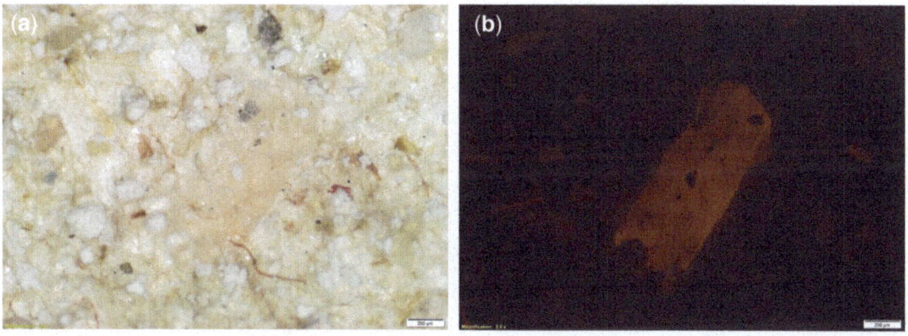

Figure 5.7 Secondary WWTP effluent extract stained with Nile Red magnified 25x, illuminated with white light (a) and a blue-green forensic light source viewed through an orange filter (b).

Differential staining with Nile Red is a promising screening tool for the visual sorting of microparticles which can reveal microplastics from background interferences. Beyond screening, this step may assist analysts in visual identification of microplastics and facilitate sorting for spectroscopic confirmation. This procedure may not stain all microplastics and some dyes used to color non-plastic microparticles may also fluoresce under the wavelengths used to excite Nile Red. Any rapid screening technique will have its limitations and must be validated by spectroscopic analysis.

5.5 DATA QUALITY OBJECTIVES (DQO)

The most commonly used analytical methods in a wastewater laboratory include those developed by the USEPA, ASTM and AWWA (Standard Methods) which routinely provide protocols to estimate data quality such as sensitivity, accuracy and reproducibility. Any standard method for determination of microplastics in secondary wastewater effluent should include Quality Control (QC) elements featured in the promulgated methods to be consistent with the established protocols and to produce data of known quality.

5.5.1 Certified reference material

In the USA (and elsewhere), US National Institute of Standards and Technology (NIST) traceable microplastic reference standards should ideally be used for quality control. Only polymer beads of limited size and composition are currently commercially available as certified reference material (CRM). This creates a challenge of sourcing representative standards. Prior to the initial validation and demonstration of capability, a laboratory should collect, extract and analyze an unfortified wastewater effluent sample to assess the types of microparticles present in the discharge of the particular WWTP. This should guide analysts in selecting a reference material that is unlikely to be detected in the final effluent, since collecting an identical duplicate sample may not always be feasible for discreet particles such as microplastics. Microbeads are a good candidate for CRM since they typically constitute less than 10% of the overall microplastic count in wastewater effluents and many are commercially available (Simon *et al.*, 2018; Ziajahromi *et al.*, 2017). Polymer films such as low density polyethylene (LDPE) and high density polyethylene (HDPE) are prevalent in wastewater and should not be used for matrix fortification.

Certified polymer microfibers are particularly challenging to source as CRM and there are currently no NIST traceable certified microplastic fibers reference standards commercially available. Fibers may contribute over 50% of the overall microplastic count in secondary wastewater effluent (Cesa *et al.*, 2017; De Falco *et al.*, 2018; Gies *et al.*, 2018; Mason *et al.*, 2016; Railo *et al.*, 2018). Some researchers have prepared microplastic fiber standards in-house to complete their own studies (Cole, 2016; Lares *et al.*, 2018). However, before a standard method can be widely adopted, it is important to have certified traceable reference standards available.

5.5.2 Initial validation and demonstration of capability

Each laboratory should verify sieve performance with a fortified reagent blank sample. This can be achieved by pouring a mixture of beads of appropriate sizes over a stack of sieves and examining fractions separately to verify the efficiency of each sieve. This procedure requires a stereomicroscope as the only analytical tool. A minimum of ten microbeads of each size should be employed. Recovery rate in excess of 90% should be accepted as satisfactory. Sieves should be periodically verified, at least once every twelve months. The laboratory should extract and analyze four laboratory fortified blanks (LFB) and one laboratory reagent blank (LRB) as part of the initial method validation. LFBs should be made using deionized reagent water fortified with a known amount of certified microplastic particles of two different sizes to be retained by different sieves (e.g., 200 and 400 μm). A minimum of ten microplastic particles of each size should be spiked. Initial acceptance criteria of 70–130% recovery and $\leq$30% RSD (Relative Standard Deviation) for an average of four results should be accepted as satisfactory. Individual laboratories may eventually develop statistically determined control limits. An LRB fortified with a mixture of beads, films and fiber microplastic reference standards would be the most accurate representation of recovery and instrument performance.

The laboratory should also demonstrate acceptable recovery rates in the wastewater matrix by extracting and analyzing a laboratory fortified matrix sample (LFM). The matrix must be fortified with a microplastic certified reference standard not typically found in wastewater. This reference standard should have two certified particle sizes (e.g. 200 and 400 μm) and initial

Table 5.3 List of QC parameters for the proposed extraction and analysis method.

QC Parameter	Acceptance Criteria	Purpose
Instrument Performance Check (IPC)	NIST library match for all microplastic standards included in the mix	Verify instrument performance each day analysis is performed
Laboratory Reagent Blank (LRB)	No microplastic particles present	Verify absence of contamination during sample preparation
Field Reagent Blank (FRB)	No microplastic particles present	Verify absence of contamination during sampling
Laboratory Fortified Blank (LFB)	70–130% recovery or in-house developed control limits	Verify accuracy of the method
Laboratory Fortified Blank (LFB) Duplicate	30% RPD (Relative Percentage Difference) or in-house developed criteria	Verify precision of the method
Laboratory Fortified Matrix (LFM)	60–140% recovery or in-house developed criteria	Evaluate method performance in the matrix during initial validation and each quarter that samples are analyzed.

acceptance criteria within 60–140% recovery should be accepted as satisfactory. Individual laboratories may develop statistically determined control limits for an LFM. The Detection Limit (DL) is defined as 1 for discrete particle analysis. Micro-FTIR or Micro-Raman must be validated by analyzing a minimum of five common microplastic polymer types. The results must match reference spectra. The laboratory may need to develop an in-house spectral library. The microplastic particle size employed for instrument validation should not exceed two times the smallest mesh size sieve used to collect samples.

Common polymer types found in wastewater include polyethylene (PE), polyethylene terephthalate (PET), polypropylene (PP), polystyrene (PS), polyamide (PA), polyvinyl chloride (PVC) and nylon. An Instrument Performance Check (IPC) containing a minimum of five common polymer types at a size no larger than 200 μm in diameter should be analyzed by Micro-FTIR or Micro-Raman each day samples are analyzed. Spectroscopic confirmation against the NIST spectral library for all polymers used in the IPC must be obtained with a match score above 80%. Each batch should include an LRB, LFB and LFB duplicate as listed in Table 5.3.

5.6 REPORTING AND DOCUMENTATION

Documentation should include sampling details, images of detected microplastic particles and spectral data. Reported data should include particle size range and particle morphology categorized as follows: fiber, film, pellet, fragment or foam. Additional useful properties that may be reported include particle color and polymer composition. Classification of individual particles by category is important for data interpretation. For example, the data can be used to identify sources of microplastic particles to the environment. To standardize reporting, the San Francisco Estuary Institute (SFEI) has submitted a microplastic-related vocabulary request that includes both operational size and particle categories to the State Water Board.

5.7 CONCLUSIONS

Method development for analysis of microplastics in wastewater effluent must include quality control and polymer confirmation to ensure that study results are comparable and verifiable. Accurate quantitation of microplastics in secondary WWTP effluent requires representative sampling and meticulous identification. Wastewater is a complex matrix containing large amounts of interferences including cellulose and fats that require effective a digestion procedure which is non-destructive to microplastics. Rapid screening by differential staining of microplastics with Nile Red is a promising and cost-effective new tool that has a potential to streamline visual sorting of microplastics. Spectroscopic confirmation is required for representative microparticles. Sampling and sample preparation steps must include quality control samples to assess method blank contamination and recovery rates of microplastics post digestion. Performance of analytical instruments such as Micro-FTIR or Micro-Raman must be verified each day analysis is performed. Furthermore, recovery rates of microplastics from the matrix must be evaluated for accuracy and precision as part of the initial method validation. An inter-laboratory study involving secondary wastewater effluents sampled from plants of different capacities is necessary to confirm the applicability of the standard method. Reporting, as a minimum, should include total volume sampled and particle characterization necessary for equivalent comparison of microplastic discharge from different WWTPs and ongoing monitoring. Challenges in sourcing representative certified reference standards and in automating spectroscopic analysis present future research needs.

ACKNOWLEDGMENTS

The authors thank the BACWA Microplastics Workgroup and San Francisco Estuary Institute for their valuable contributions.

REFERENCES

Andrady A. L. (2011). Microplastics in the marine environment. *Marine Pollution Bulletin*, **62**(8), 1596–605.

Araujo C. F., Nolasco M. M., Ribeiro A. M. P. and Ribeiro-Claro P. J. A. (2018). Identification of microplastics using Raman spectroscopy: latest developments and future prospects. *Water Research*, **142**, 426–440.

Carr S. A., Liu J. and Tesoro A. G. (2016). Transport and fate of microplastic particles in wastewater treatment plants. *Water Research*, **91**, 174–82.

Cesa F. S., Turra A. and Baruque-Ramos J. (2017). Synthetic fibers as microplastics in the marine environment: a review from textile perspective with a focus on domestic washings. *Science of the Total Environment*, **598**, 1116–112.

Cole M. (2016). A novel method for preparing microplastic fibers. *Scientific Reports (Nature Research)*, **6**, 34519.

De Falco F., Gullo M. P., Gentile G., Di Pace E., Cocca M., Gelabert L., Brouta-Agnésa M., Rovira A., Escudero R., Villalba R., Mossotti R., Montarsolo A., Gavignano S., Tonin C. and Avella M. (2018). Evaluation of microplastic release caused by textile washing processes of synthetic fabrics. *Environmental Pollution*, **236**, 916–925.

Dris R., Gasperi J., Rocher V., Saad M., Renault N. and Tassin B. (2015). *Environmental Chemistry*, **12**(5), 592–599.

Dyachenko A., Mitchell J. and Arsem N. (2017). Extraction and identification of microplastic particles from secondary wastewater treatment plant (WWTP) effluent. *Analytical Methods*, **9**, 1412–1418.

Gago J., Carretero O., Filgueiras A. V. and Viña S. L. (2018). Synthetic microfibers in the marine environment: a review on their occurrence in seawater and sediments. *Marine Pollution Bulletin*, **127**, 365–376.

Gies E. A., LeNoble J. L., Noël M., Etemadifar A., Bishay F., Hall E. R. and Ross P. S. (2018). Retention of microplastics in a major secondary wastewater treatment plant in Vancouver, Canada. *Marine Pollution Bulletin*, **133**, 553–561.

Hanvey J., Lewis P., Lavers J., Crosbie N., Posa K. and Clarke B. (2016). A review of analytical techniques for quantifying microplastics in sediments. *Analytical Methods*, **9**, 1361–1368.

Lares M., Ncibi M. C., Sillanpää M. and Sillanpää M. (2018). Occurrence, identification and removal of microplastic particles and fibers in conventional activated sludge process and advanced MBR technology. *Water Research*, **133**, 236–246.

Li J., Liu H. and Paul Chen J. (2018). Microplastics in freshwater systems: a review on occurrence, environmental effects, and methods for microplastics detection. *Water Research*, **137**, 362–374.

Löder M. G. J., Imhof H. K., Ladehoff M., Löschel L. A., Lorenz C., Mintenig S., Piehl S., Primpke S., Schrank I., Laforsch C. and Gerdts G. (2017). Enzymatic purification of microplastics in environmental samples. *Environmental Science & Technology*, **51**(24), 14283–14292.

Maes T., Jessop R., Wellner N., Haupt K. and Mayes A. G. (2017). A rapid-screening approach to detect and quantify microplastics based on fluorescent tagging with Nile Red. *Scientific Reports*, **7**, 44501.

Mason S. A., Garneau D., Sutton R., Chu Y., Ehmann K., Barnes J., Fink P., Papazissimos D. and Rogers D. L. (2016). Microplastic pollution is widely detected in US municipal wastewater treatment plant effluent. *Environmental Pollution*, **218**, 1045–1054.

Michielssen M. R., Michielssen E. R., Niac J. and Duhaime M. B. (2016). Fate of microplastics and other small anthropogenic litter (SAL) in wastewater treatment plants depends on unit processes employed. *Environmental Science: Water Research & Technology*, **2**, 1064–1073.

Mintenig S. M., Int-Veen I., Löder M. G. J., Primpke S. and Gerdts G. (2017). Identification of microplastic in effluents of waste water treatment plants using focal plane array-based micro-Fourier-transform infrared imaging. *Water Research*, **08**, 365–372.

Munno K., Helm P. A., Jackson D. A., Rochman C. and Sims A. (2018). Impacts of temperature and selected chemical digestion methods on microplastic particles. *Environmental Toxicology and Chemistry*, **37**(1), 91–98.

Murphy F., Ewins C., Carbonnier F. and Quinn B. (2016). Wastewater Treatment Works (WwTW) as a source of microplastics in the aquatic environment. *Environmental Science & Technology*, **50**(11), 5800–8.

Primpke S., Lorenz C., Rascher-Friesenhausenbc R. and Gerdtsa G. (2017). An automated approach for microplastics analysis using focal plane array (FPA) FTIR microscopy and image analysis. *Analytical Methods*, **9**, 1499–1511.

Railo S., Talvitie J., Setälä O., Koistinen A. and Lehtiniemi M. (2018). Application of an enzyme digestion method reveals microlitter in Mytilus trossulus at a wastewater discharge area. *Marine Pollution Bulletin*, **130**, 206–214.

Sedlak M., Sutton R., Box C., Sun J. and Lin D. (2017). Sampling and Analysis Plan for Microplastic Monitoring in San Francisco Bay and Adjacent National Marine Sanctuaries. *San Francisco Estuary Institute (SFEI) Contribution*. No.819. SFEI, Richmond, CA.

Shim W. J., Song Y. K., Hong S. H. and Jang M. (2016). Identification and quantification of microplastics using Nile Red staining. *Marine Pollution Bulletin*, **113**(1–2), 469–476.

Silva A. B., Bastos A. S., Justino C. I. L., da Costa J. P., Duarte A. C. and Rocha-Santos T. A. P. (2018). Microplastics in the environment: challenges in analytical chemistry-A review. *Analytica Chimica Acta*, **1017**, 1–19.

Simon M., van Alst N. and Vollertsen J. (2018). Quantification of microplastic mass and removal rates at wastewater treatment plants applying Focal Plane Array (FPA)-based Fourier Transform Infrared (FT-IR) imaging. *Water Research*, **142**, 1–9.

Sutton R., Mason S. A., Stanek S. K., Willis-Norton E., Wren I. F. and Box C. (2016). Microplastic contamination in the San Francisco Bay, California, USA. *Marine Pollution Bulletin*, **109**(1), 230–235.

Tagg A. S., Sapp M., Harrison J. P. and Ojeda J. J. (2015). Identification and quantification of microplastics in wastewater using focal plane array-based reflectance Micro-FT-IR imaging. *Analytical Chemistry*, **87**(12), 6032–40.

Talvitie J., Heinonen M., Pääkkönen J. P., Vahtera E., Mikola A., Setälä O. and Vahala R. (2015). Do wastewater treatment plants act as a potential point source of microplastics? Preliminary study in the coastal Gulf of Finland, Baltic Sea. *Water Science & Technology*, **2**(9), 1495–504.

Talvitie J., Mikola A., Setälä O., Heinonen M. and Koistinen A. (2017). How well is microlitter purified from wastewater? A detailed study on the stepwise removal of microlitter in a tertiary level wastewater treatment plant. *Water Research*, **109**, 164–172.

Wang Z., Su B., Xu X., Di D., Huang H., Mei K., Dahlgren R. A., Zhang M. and Shang X. (2018). Preferential accumulation of small (<300 μm) microplastics in the sediments of a coastal plain river network in eastern China. *Water Research*, **144**, 393–401.

Zhang K., Shi H., Peng J., Wang Y., Xiong X., Wu C. and Lam P. K. S. (2018). Microplastic pollution in China's inland water systems: a review of findings, methods, characteristics, effects, and management. *Science of the Total Environment*, **630**, 1641–1653.

Ziajahromi S., Neale P. A., Rintoul L. and Leusch F. D. (2017). Wastewater treatment plants as a pathway for microplastics: development of a new approach to sample wastewater-based microplastics. *Water Research*, **112**, 93–99.

Chapter 6

Microplastics in sewage sludge: Captured but released?

A. L. Lusher, R. R. Hurley and C. Vogelsang

Norwegian Institute for Water Research, Oslo, Norway

Keywords: Agriculture, Contamination, Effluents, Microlitter, Nanoplastic, Plastic, Wastewater, Wastewater Treatment Plants

6.1 INTRODUCTION

Sewage sludge is a solid by-product of the wastewater treatment process. Global efforts to improve the treatment of wastewater, coupled with population growth, is leading to an increase in the volume of sludge production. Conventionally, sludge has been disposed of through landfilling; however, legislation, particularly in developed nations, has significantly decreased this practice over the past 25 years (e.g. EU Council Directive 86/278/EEC) (EU, 1986). Instead, strategies for valorising or repurposing sludge for 'beneficial use' have been promoted, with a recent interest in approaches that fit within a circular economic framework (Kacprzak *et al.*, 2017). Several mechanisms for disposal exist, including application to land as soil amendment and incineration. However, wide disparity in the handling and disposal of sewage sludge exists across the globe.

Wastewater treatment plants (WWTPs) receive plastic particles from a wide range of sources including domestic, urban and industrial effluents. These particles exist on a broad size spectrum from large litter items, to micro- and nano-sized particles. Anthropogenic particles are either captured during different stages of wastewater treatment or are released in the final effluent. Several studies

Editors: Hrissi K. Karapanagioti and Ioannis K. Kalavrouziotis
doi: 10.2166/9781789061697_0085

have pointed towards high retention of microplastics (particles < 5 mm; GESAMP, 2015) by WWTPs, although trapping efficiencies vary significantly between plants. Particles that are captured during certain treatment stages are likely to be transferred to the sludge phase, preventing them from being released in the final effluent. Nevertheless, due to practices employed for dealing with sewage sludge, release into the environment may still occur. Sludge may be disposed of through several mechanisms including improper dumping, landfilling, incineration or land application. Some disposal techniques represent a potential release of very large numbers of microplastics to the environment. Recent focus has been directed towards the sludge fraction based upon high concentrations of particles observed in sludge and its reuse in agricultural practices in many countries. WWTPs are now considered to represent an important pathway for microplastic contamination, as significant amounts of microplastics are estimated to be released into the terrestrial environment via sludge application. This highlights an urgent need for a thorough understanding of the fate and impacts of sludge-associated microplastics. In this chapter we discuss the treatment processes that may lead to the capture and transfer of microplastics to sludge, the mechanisms through which sludge disposal or re-use may constitute a direct release of microplastics to the environment, and the potential consequences of this release.

6.2 TRANSFER OF MICROPLASTICS TO THE SLUDGE PHASE

For microplastics to end up in sludge they first need to be physically removed from the water phase and then survive the different stages of sludge treatment. Existing studies conducted in full-scale WWTPs suggest that commonly applied treatment processes are rather efficient barriers to microplastics, at least for the fraction of microplastic particles that have been studied. The overall removal ratio is typically in the range of 84–99.9% for WWTPs applying either biological treatment, chemical precipitation or both (Table 6.1). A direct comparison of the observed removals at different WWTPs is somewhat hampered by many underlying differences in study approaches. These include sampling methods used (e.g. sampling duration, volumes, particles size detection limits), sample preparation, particle quantification and polymer verification. In addition, site-specific conditions during sampling campaigns can contribute to underlying differences including treatment equipment applied, operating conditions of individual WWTPs, and time- or situation-specific conditions, such as operational stabilities and peaking loads. Nevertheless, the probability for a given microplastic particle to be captured by a given treatment process may be predicted from specific properties of that particle, inherent mechanisms that are applied by the treatment unit to remove particles from the water phase and concurrent operational conditions. Particle size, shape, specific density and surface characteristics (roughness, charge, hydrophilicity/hydrophobicity) are the most important properties when considering potential removal during treatment.

6.2.1 Coarse bar screens and grit chambers

Coarse materials, such as sticks, rags and other debris that could damage and clog processing equipment and reduce overall reliability and effectiveness, are removed by screening through a coarse bar screen. Even if there is typically $\geq$6 mm between bars, accumulation of material will increase retention of smaller particles, such as those within the size range of large microplastics (1–5 mm). Hence, trapping efficiency is dependent on both the amount of accumulated matter and cleaning frequency. A grit chamber is commonly placed after coarse bar screens to remove heavy, small-sized mineral particles (c. $>$0.1 mm), such as gravel and sand. Grit chambers are typically sized to remove particles with specific gravities of around 2.65, but analysis of grit removal data indicates specific gravity can range from 1.3–2.7 (WPCF, 1985), which may then include the higher density fraction of microplastics. For example, Michielssen *et al.* (2016) found that between 35–59% of microplastics (20–4750 µm) were removed by preliminary treatment units (bar screens and grit chambers). Approximately 58% of microplastics were removed by an aerated grit chamber in Beijing, China (Yang *et al.*, 2019a). Furthermore, Carr *et al.* (2016) estimated that $\sim 7.78 \times 10^6$ microplastics could be collected in a grit chamber during daily production, although this was based on a single particle count in a 2.1 g grit sample. It should be noted that, usually, none of the material collected at this stage will end up in the final sludge, as they are typically incinerated or landfilled.

6.2.2 Grease skimmers

Grease in incoming wastewater is usually removed in the grit chamber or in the primary settler by surface skimming. Since many grit chambers apply aeration to rinse grit particles free of organic matter (thereby increasing their specific gravity), microplastic particles with a density lower than the water itself (i.e. densities <1.0 g cm^{-3}) are given an additional lift to the surface. Murphy *et al.* (2016) observed 45% removal (by particle count) of microplastics (63–5000 µm) by coarse screens (6–19 mm) and an aerated grit chamber fitted with a grease skimmer, with the highest concentration found in collected grease. Grease samples (2.5 g) contained on average c. 20 microplastic particles with polyethylene, polyester, polyethylene terephthalate and alkyd most commonly observed (Murphy *et al.*, 2016). Similarly, 24 particles (dominated by blue polyethylene fragments) were identified from a primary skimmer (sample size: 5 g) (Carr *et al.*, 2016). However, based on total grease volume collected daily, only about 6% of microparticles appeared to be contributed by the grease skimmer, based on our calculations.

Table 6.1 Summary of studies that investigate microplastic contamination in sewage sludge.

Location	WWTP Type (Population Equivalent)	Reported Trapping Efficiency	Treatment Type	Mean Sludge MP Concentrations	MP Size Range Analysed (μm)	Dominant MP Types Shape	Dominant MP Types Polymer	Study
Italy	Tertiary (1.2 million)	84%	NR	113 particles g^{-1} (DW)	10–5,000	Films	Acrylonitrile-butadiene	Magni *et al.* (2019)
Australia	NR (NR)	NR	Aerobic & anaerobic digestion	966 microbeads kg^{-1}	<1,000	Only microbeads investigated	NR	Wijesekara *et al.* (2018)
China	28 WWTPs: Mixed types (51,900–705,000)	NR	Various	1,565–56,386 particles kg^{-1} (DW)	37–5,000	Fibres	Polyolefin	Li *et al.* (2018)
China	NR (NR)	NR	NR	240.3 ± 31.4 particles g^{-1} (DW)	60–4,200	Fragments	Nylon	Liu *et al.* (2019)
Korea	3 treatment plants (A: 67,700) (B: 235,700) (C: 245,200)	98%	A, B: Sludge thickening and dehydration C: Thickening, anaerobic digestion and dehydrator	A: 14.9 particles g^{-1} (DW) B: 9.6 particles g^{-1} (DW) C: 13.2 particles g^{-1} (DW)	106–5,000	Fragments	NR	Lee and Kim (2018)
Finland	Tertiary (55,000)	98%	Anaerobic digestion (and dewatering)	4.2–28.7 particles g^{-1} (DW)	250–5,000	Fibres	Polyester	Lares *et al.* (2018)
Canada	Secondary (NR)	98%	NR	14.9 (primary) & 4.4 (secondary) suspected particles g^{-1} (DW)	<5,000	Fibres	NCR	Gies *et al.* (2018)
Norway	2 tertiary, 4 secondary, 2 primary (18,150–615,000)	NR	Various	1,701–19,837 particles kg^{-1} (DW)	50–5,000	Beads	Polyethylene	Lusher *et al.* (2018)
Netherlands	1 tertiary, 1 secondary WWTP, 1 STW (NR)	72%	NR	370–950 particles kg^{-1} (WW)	10–5,000	NR	NR	Leslie *et al.* (2017)

Germany	1 tertiary, 5 secondary (11,000–56,000)	NR	NR	1,000–24,000 particles kg^{-1} (DW)	<500	NR	Polyethylene	Mintenig *et al.* (2017)
Germany	STW (50,000)	NR	NR	223 particles 100 mL^{-1} or 495 particles g^{-1} (DW)	0.48–500	NR	NR	Sujathan *et al.* (2017)
Finland	Tertiary (800,000)	>99%	Excess and dried sludge; anaerobic digestion & dewatering	76.3 (excess) & 186.7 (dried) microlitter particles g^{-1}	100–5,000	NR	NR	Talvitie *et al.* (2017)
Denmark	5 WWTPs (NR)	92.6%–99.7%	NR	169,000 particles g^{-1}	20–500	NR	Polyethylene	Vollertsen and Hansen (2017)
Spain	Primary (210,000)	NR	Primary and anaerobically-digested sludge	NR: MPs present	NR	NR	NR	Bayo *et al.* (2016)
Scotland	Secondary (650,000)	98.4%	Dewatered sludge cake which is produced after sludge thickening and centrifugation	Approx. 2.5 particles 2.5 g^{-1}	NR	NR	Polyester	Murphy *et al.* (2016)
Ireland	7 WWTPs (6,500–2.4 m)	NR	Various	4,196–15,385 particles kg^{-1} (DW)	250–5,000	Fibres	HDPE	Mahon *et al.* (2017)
USA	NR	NR	Dewatering & anaerobic digestion	4 fibres g^{-1}	NR	Only fibres investigated	NCR	Zubris and Richards (2005)
USA	7 tertiary, 1 secondary (NR)	NR	Return activated sludge	1 particle 20 mL^{-1}	45–5,000	NR	NR	Carr *et al.* (2016)
Sweden	Tertiary (14,000)	>99.9%	Partially dewatered	16,700 particles kg^{-1} (DW)	300–5,000	Fibres	NCR	Magnusson and Norén (2014)

NR: Not reported; NCR: No characterisation reported; DW: Dry weight; WW: wet weight.

Collected grease is often treated together with sludge from other treatment units at a WWTP. Therefore, grease and foam skimmers may be important vectors for the transfer of low-density microplastics to the sludge treatment line. Our study in Norway identified that 62% of particles extracted from sludge had a specific density lower than that of pure water (Lusher *et al.*, 2018), which could suggest these floating particles may have been removed during this early separation phase.

6.2.3 Primary and secondary clarifiers

Higher-density microplastic particles may be separated from wastewater by settling in primary and/or secondary clarifiers. Primary clarifiers remove readily settleable solids and floating material (if grit chambers are not equipped with grease skimmers), thus reducing suspended solid content. Michielssen *et al.* (2016) found that 62–82% of residual microplastic particles (>20 μm) were removed by the primary clarifier unit. This appears to agree with the 61% removal (post-grease skimming) observed by Murphy *et al.* (2016), though low dosages of a coagulant and a flocculant were used to improve particulate matter removal in the primary clarifier at that WWTP.

Secondary clarifiers clarify biologically and/or chemically treated wastewater and concentrate sludge before it is sent for thickening or returned to the bioreactor (in activated sludge processes). Naturally occurring biological floc formation, or chemical coagulation and flocculation processes, significantly improve settling rates of particulate matter. Moreover, secondary clarifiers are typically designed with somewhat longer settling times than primary settlers to provide enough time for clarification. While Murphy *et al.* (2016) observed that only 7% of residual microplastics were found in the effluent from the secondary clarifier (i.e. 93% removal), the observed removal over the secondary treatment step (activated sludge treatment and secondary clarifier) at the two WWTPs in Michielssen *et al.*'s (2016) study was a mere 12–61%. Coagulants, and particularly flocculants, are used to improve the settling characteristics of sludge flocs, thus apparently also improving the removal of residual microplastics.

6.2.4 Other treatment steps

Membrane bioreactors (MBRs) are increasingly popular at full-scale WWTPs to separate activated sludge and treated water since the secondary clarifier can then be omitted. An MBR can be expected to provide significantly improved retention of microplastics (Lares *et al.*, 2018).

Some WWTPs apply a polishing step, often a rapid sand filter, to capture sludge flocs that may escape the secondary clarifier. Particle characteristics (e.g. size distribution and surface properties), the volumetric loading and ageing of the filter, as well as the type of sand used and the grain size distribution determine removal efficiency. Typically, about 90% of 10 μm particles are removed, whereas only around 10% of 2 μm particles are removed (Tchobanoglous *et al.*, 2003).

All collected sludge, except that from the preliminary treatment units (collected solids, such as sticks, rags and sand) are mixed and treated together. Treatment is performed to reduce sludge weight, volume and potential health risks. The sludge treatment process varies considerably between treatment plants and may involve several stages: thickening, stabilisation through lime addition, aerobic and/or anaerobic digestion and composting, conditioning, dewatering and heat drying. Typically, this depends on the method of sludge disposal being used and some sludge may therefore not go through stabilisation processes.

Such processes could contribute to the fragmentation and breakdown of plastic particles. Studies have indicated potential fragmentation of e.g. nylon fibres and polyethylene fragments by shredding and flaking during lime stabilisation resulting from the combination of high pH and mechanical mixing as well as melting, and blistering of high-density polyethylene and polyethylene fragments during heat drying (Cole *et al.*, 2013; Mahon *et al.*, 2017; Zubris & Richards, 2005). Importantly, the reject water from dewatered sludge may contain a large fraction of the microplastic particles; Murphy *et al.* (2016) found that 20% of microplastic particles were returned to the WWTP inlet by the reject water.

6.3 REPORTED CONCENTRATIONS OF MICROPLASTICS IN SEWAGE SLUDGE

At the time of writing, 19 published studies report concentrations of microplastics in sewage sludge (Table 6.1). Several different methods are used in the collection, treatment and analysis of sludge samples making comparisons difficult. It is also important to note that the term sludge, as referred to in these studies, has the potential to differ somewhat regarding its definition. Several studies investigate concentrations in sludge taken from specific treatment steps (e.g. return activated sludge) as opposed to the final sludge. Additionally, concentrations are reported in some instances prior to, or following, dewatering or sludge treatment. All these factors are likely to influence sludge concentrations, and so it is not appropriate to perform a direct comparison of the results. Comparison is further hampered by methodological disparities that exist due to an absence of standardised or harmonised analytical methodologies.

6.3.1 Methods used

Sludge samples are often collected with grab samplers and can be collected as single samples or over a period of time to form composites (e.g. Lusher *et al.*, 2018). Sludge is composed of a complex mixture of solid material including microorganisms, organic and inorganic materials, and microplastics. Several different methods to break down complex and organic-rich samples have been proposed including density separation (using NaCl, NaI, $ZnCl_2$, $BaCl_2$), peroxide treatment (H_2O_2), treatment with sodium hydroxide (NaOH), or sieving and visual sorting. Following rigorous testing and recovery tests with these methods, Fenton's reagent has been shown to be efficient for removing organic matter and preserving plastic particles (Hurley *et al.*, 2018). Enzymatic digestion is also effective at removing organic material, including resistant fractions such as cellulose; it is more expensive and time consuming than other approaches but may represent the optimum technique for analysing very small microplastic particles (e.g. <50 μm) or particularly complex sludge samples. The significant potential for variability in sludge samples resulting from their content (i.e. sources of solid material entering the plant) and subsequent treatment (which may influence different organic fractions differently and alter factors such as sludge pH) hinder efforts to establish a more harmonised methodological approach. To counteract this, innovative approaches such as the development of rare element-doped micro- and nanoplastic particles have been implemented, which may be effective at robustly evaluating the influence of different water or sludge treatment steps on plastic concentrations for different particle types (Mitrano *et al.*, 2019; Schmiedgruber *et al.*, 2019). Ongoing methodological optimisation is required to increase the resolution of analyses and establish an holistic understanding of microplastic dynamics both within WWTPs and associated with the fate of sewage sludge.

6.3.2 Comparisons between studies

One of the key difficulties when discussing concentrations of microplastics in sewage sludge relates to the fact that no two WWTPs are exactly alike. This hampers comparisons between different studies and complicates investigations into the influence of different WWTP characteristics, such as the population equivalent, influent sources, or the treatment processes that are employed. Since microplastic particles accumulate in the mixed sludge, and this is largely recycled to provide enough sludge for the active biomass, a direct comparison between influent and effluent should be conducted under thorough research conditions.

Despite this, some broad points of note can be gleaned from the current literature. For example, concentrations are higher when smaller size fractions are included in the analysis (i.e. down to 10–20 μm) (Magni *et al.*, 2019; Sujathan *et al.*, 2017; Vollertsen & Hansen, 2017). It is apparent from Table 6.1 that too much disparity exists between studies to establish any clear-cut controls on sludge microplastic concentrations, such as the effect of sludge treatment processes. However, within studies we can see potential trends. Specific differences in the level of treatment is related to specific plant characteristics. These alterations in shape and size suggest that sludge treatment processes may influence microplastic particles and that WWTPs may be increasing the proportion of particles in smaller size classes. Hence, whilst comparisons between studies are not possible, the current literature presents some important findings that warrant further research.

6.4 FATE OF MICROPLASTICS IN SEWAGE SLUDGE

With the gathering evidence of high concentrations of microplastics in sewage sludge, concern amongst scientists has led to calls to understand the consequences of current disposal and reuse practices. The fate of microplastics contained within sewage sludge is tied to the practices and legislation employed for sludge disposal and there are two main practices employed that have the potential to either partly or completely limit release: thermal treatment and landfilling. Thermal treatment refers to a range of techniques that are likely to destroy microplastic particles, including incineration, pyrolysis and gasification, and co-incineration (Kacprzak *et al.*, 2017), and represent technologies that utilise the energy stored within sludge for generation of heat or, ultimately, power. The principle of thermal treatment is based upon the combustion, and thus destruction, of the organic components, which will likely include microplastics if combustion temperatures are significantly high (c. 400–550°C). Thermal treatment is typically performed along with phosphorus recovery and flue gas scrubbing in higher income countries, which is currently expensive (Fytili & Zabaniotou, 2008). Incineration still occurs in less-economically developed regions, albeit with fewer controls on emissions or particle losses.

Landfilling also represents a means of disposal whereby microplastic release is expected to be limited. Where landfilling is performed appropriately, particles will be trapped. Nonetheless, this approach for managing sewage sludge is rapidly declining due to legislative discouragement, capacity limitations and a lack of public acceptance (Milieu Ltd. *et al.*, 2013). Additionally, there has not yet been detailed research on the release of microplastics in landfill leachate or the release of particles resulting from improper practice.

Several countries have strict legislation on the proper disposal of sludge. In contrast, dumping of sludge on land still occurs in some cases. For example, more than 80% of sludge in China is disposed of by improper dumping (Yang *et al.*, 2015). This is expected to act as a significant source of microplastics to the surrounding environment, where the mechanisms of release are mediated by the manner of dumping. Factors such as proximity to aquatic systems, aspect and slope, and local geomorphology will play a role in the spreading of microplastics from dump sites. The potential magnitude of this release is difficult to estimate where surface dumping is not regulated or documented. Moreover, the concentrations of microplastics in sludge may be tied to the connectivity of potential sources to WWTPs (e.g. industrial effluents), the degree of wastewater treatment, and social practices such as domestic clothes washing (Napper & Thompson, 2016).

The final disposal mechanism for sludge discussed here is reuse as a soil conditioner or for fertilisation. All sludge first requires to be stabilised and hygienised before it is used as soil conditioner. Composting is one way of doing this, whilst anaerobic digestion and/or lime stabilisation are other common options. Following these processes, the material is often directly applied to land for use in soil production. Biosolids are heavily regulated and application is sometimes restricted in the US, although federal legislation is related to the varying levels of contaminants (and microplastics are currently not considered a contaminant) (Harrison *et al.*, 1999). Conversely, the practice of biosolid application to land is currently encouraged by legislation in many countries, including in the EU (Council Directive 86/278/EEC; see EU (1986)) due to the nutrient rich properties of biosolids.

Approximately 40% of sludge is used for agricultural land application in the EU (Fytili & Zabaniotou, 2008; Mateo-Sagasta *et al.*, 2015), although this varies widely amongst member states. It is expected that this represents a major release pathway for microplastics into the environment. Nizzetto *et al.* (2016) estimate that between 43,000–630,000 tons of microplastics are added to European farmland each year through sludge application, and Lusher *et al.* (2018) calculated that approximately 584 billion (584×10^9) microplastic particles are released to the Norwegian environment in one year via sludge application as a soil conditioner. Once applied to fields, microplastics may become incorporated into soil (Rillig *et al.*, 2017). Environmental processes could further encourage the movement of particles from the terrestrial system into water ways and eventually the oceans. This has the effect of propagating contamination across a wider area and potentially contaminating systems otherwise free from microplastics (Hurley & Nizzetto, 2018).

6.4.1 Potential implications associated with microplastic release

Only a small number of studies have thus far investigated interactions between microplastics and terrestrial organisms. Reported adverse effects include reduction in growth and reproduction in Collembola (Zhu *et al.*, 2018), histopathological damage in earthworm guts (Rodriguez-Seijo *et al.*, 2017), and transfer of polybrominated diphenyl ethers (PBDEs) to earthworms (Gaylor *et al.*, 2013). However, other studies report no effects (e.g. Jemec Kokalj *et al.*, 2018) or only demonstrate negative impacts at concentrations beyond environmentally relevant levels (Cao *et al.*, 2017; Huerta Lwanga *et al.*, 2016). Whilst it is important to consider that microplastics may accumulate and persist in soils through successive sludge application, it is evident from the current literature that the potential impacts of microplastics to soil biota are not yet clear.

Several studies have demonstrated that microplastics can act as a vector for contaminants, such as metals (e.g. Brennecke *et al.*, 2016) or persistent organic pollutants (POPs) (e.g. Bakir *et al.*, 2014). WWTPs concentrate contaminants, and microplastics contained within the sludge phase may become enriched due to sorption to the particle surface or development of a biofouling layer (e.g. Wijesekara *et al.*, 2018). Once released, microplastics from sludge may carry these contaminants into different environmental compartments. It is important, however, to place the evidence for the risks posed by microplastics into perspective. For example, studies have pointed towards the role of microplastics in the spread of antibiotic resistance (Arias-Andres *et al.*, 2018; Eckert *et al.*, 2018). This is highlighted as an important factor in what is a major global challenge. Despite this, the spread of antibiotic resistance occurs at a much larger scale associated with the organic component of the sludge (Bondarczuk *et al.*, 2016; Chen *et al.*, 2016), whereas the transfer associated with microplastic particles is likely to represent a very small proportion. Indeed, for sorbed contaminants – such as antibiotics – there is typically a much higher affinity for organic matter than for microplastic particles (Xu *et al.*, 2018) and the transfer of contaminants from microplastics is significantly outstripped by natural pathways (Koelmans *et al.*, 2016). Despite this, microplastics persist for longer and have the potential to spread across wider spatial scales. Hence, microplastics have been described as 'reservoirs' for antibiotic resistance in the marine environment (Yang *et al.*, 2019b). So, the role of microplastic particles must be appropriately evaluated against a range of conditions.

6.5 CONCLUSIONS

A large proportion of the microplastics captured by WWTPs are likely to be transferred to the sludge phase, leading to the high concentrations of microplastics observed in this matrix type. There are several mechanisms through which these particles may be released to the environment, in which sewage sludge may represent a highly significant release pathway. Legislation to limit the quantity of microplastics found in sludge is difficult when WWTPs have limited control over the sources of microplastics entering the plant and in cases where there is no technology currently available to capture microplastics and separate them from the sludge. Norway's current Fertiliser Product Regulations (Gjødselvareforskriften) state that 'the total content of plastic, glass or metal particles with a size greater than 4 mm shall not exceed 0.5% by weight of the total dry matter' LMD KLD and HOD (2019) (translated from the Norwegian). Measures to prevent microplastic accumulation in sludge should be implemented upstream of the sludge process, avoiding or minimising the transfer of microplastics to the sludge treatment line. Available studies have shown that a fraction of microplastics are captured by simple treatment such as coarse bar screens, sand traps and grease skimmers (maybe as much as 80% according to Murphy *et al.*, 2016). If the material collected by the grease skimmers are incinerated or similarly disposed, more of the microplastics may be prevented from ending up in the final sludge. However, the grease is also a valuable substrate for biogas production during anaerobic sludge digestion. A much deeper holistic approach to sludge production and microplastic capture is required.

Instead, efforts should focus on curbing the input of microplastics to WWTPs through effective source control. Intentionally added microplastics, including those found in personal care and household products, are currently being reviewed under a proposed EU 'Registration, Evaluation, Authorisation and Restriction of Chemicals' (REACH) restriction. The European Chemicals Agency (ECHA) recently highlighted that sludge likely represents the main release pathway for these microplastic types (ECHA, 2018). Reducing these sources should reduce observed sludge concentrations but other microplastic types are important too. Fibres and fragments often dominate in the sludge and the sources of these particles must also be addressed.

Additional research is urgently required to investigate the controls, fate and impacts of sludge-associated microplastics. Future research should address the real risks posed by microplastics in sludge and establish threshold effects levels. It is important to quantify sludge as a source of microplastics to the environment, including in developing regions. Finally, solutions to reduce the sources of microplastics to WWTPs should be assessed to effectively reduce the microplastic content of sewage sludge.

ACKNOWLEDGMENTS

The authors would like to thank Dr Luca Nizzetto for his comments and earlier discussions concerning this chapter.

REFERENCES

Arias-Andres M., Klümper U., Rojas-Jimenez K. and Grossart H.-P. (2018). Microplastic pollution increases gene exchange in aquatic ecosystems. *Environmental Pollution*, **237**, 253–261.

Bakir A., Rowland S. J. and Thompson R. C. (2014). Enhanced desorption of persistent organic pollutants from microplastics under simulated physiological conditions. *Environmental Pollution*, **185**, 16–23.

Bayo J., Olmos S., López-Castellanos J. and Alcolea A. (2016). Microplastics and microfibers in the sludge of a municipal wastewater treatment plant. *International Journal of Sustainable Development and Planning*, **11**, 812–821.

Bondarczuk K., Markowicz A. and Piotrowska-Seget Z. (2016). The urgent need for risk assessment on the antibiotic resistance spread via sewage sludge land application. *Environment International*, **87**, 49–55.

Brennecke D., Duarte B., Paiva F., Caçador I. and Canning-Clode J. (2016). Microplastics as vector for heavy metal contamination from the marine environment. *Estuarine, Coastal and Shelf Science*, **178**, 189–195.

Cao D., Wang X., Luo X., Liu G. and Zheng H. (2017). Effects of polystyrene microplastics on the fitness of earthworms in an agricultural soil IOP Conf. Ser. *Earth Environmental Science*, **61**, 012148.

Carr S. A., Liu J. and Tesoro A. G. (2016). Transport and fate of microplastic particles in wastewater treatment plants. *Water Research*, **91**, 174–182.

Chen Q., An X., Li H., Su J., Ma Y. and Zhu Y.-G. (2016). Long-term field application of sewage sludge increases the abundance of antibiotic resistance genes in soil. *Environment International*, **92**–93, 1–10.

Cole M., Lindeque P., Fileman E., Halsband C., Goodhead R., Moger J. and Galloway T. S. (2013). Microplastic Ingestion by Zooplankton. *Environmental Science & Technology*, **47**(12), 6646–6655.

Eckert E. M., Di Cesare A., Kettner M. T., Arias-Andres M., Fontaneto D., Grossart H.-P. and Corno G. (2018). Microplastics increase impact of treated wastewater on freshwater microbial community. *Environmental Pollution*, **234**, 495–502.

European Council (EU) (1986). EU Council Directive of 12 June 1986 on the Protection of the Environment, and in Particular of the Soil, when Sewage Sludge is Used in Agriculture (86/278/EEC). Council of the European Communities. Official Journal of the European Communities *No. L*, 181/6–12. See: https://eur-lex.europa.eu/eli/dir/1986/278/oj (accessed May 2019).

European Chemicals Agency (ECHA) (2018). Intentionally added microplastics likely to accumulate in terrestrial and freshwater environments. ECHA/PR/18/15 See: https://echa.europa.eu/-/intentionally-added-microplastics-likely-to-accumulate-in-terrestrial-and-freshwater-environments (accessed May 2019).

Fytili D. and Zabaniotou A. (2008). Utilization of sewage sludge in EU application of old and new methods: a review. *Renewable & Sustainable Energy Reviews*, **12**, 116–140.

Gaylor M. O., Harvey E. and Hale R. C. (2013). Polybrominated Diphenyl Ether (PBDE) Accumulation by Earthworms (*Eisenia fetida*) Exposed to Biosolids-, Polyurethane Foam Microparticle-, and Penta-BDE-Amended Soils. *Environmental Science & Technology*, **47**, 13831–13839.

GESAMP (2015). Chapter 3.1.2 Defining 'microplastics'. Sources, Fate and Effects of Microplastics in the Marine Environment: A Global Assessment. In: (IMO/FAO/UNESCO–IOC/UNIDO/ WMO/IAEA/UN/
UNEP/UNDP Joint Group of Experts on the Scientific Aspects of Marine Environmental Protection (GESAMP)), P. J. Kershaw (ed.), Rep. Stud. GESAMP No. 90, 96p, London.

Gies E. A., LeNoble J. L., Noël M., Etemadifar A., Bishay F., Hall E. R. and Ross P. S. (2018). Retention of microplastics in a major secondary wastewater treatment plant in Vancouver, Canada. *Marine Pollution Bulletin*, **133**, 553–561.

Harrison E. Z., McBride M. B. and Bouldin D. R. (1999). Land application of sewage sludges: an appraisal of the US regulations. *International Journal of Environment and Pollution*, **11**(1), 1–36.

Huerta Lwanga E., Gertsen H., Gooren H., Peters P., Salánki T., van der Ploeg M., Besseling E., Koelmans A. A. and Geissen V. (2016). Microplastics in the terrestrial ecosystem: implications for lumbricus terrestris (Oligochaeta, Lumbricidae). *Environmental Science & Technology*, **50**, 2685–2691.

Hurley R. R. and Nizzetto L. (2018). Fate and occurrence of micro(nano)plastics in soils: knowledge gaps and possible risks. *Current Opinion in Environmental Science & Health*, **1**, 6–11.

Hurley R. R., Lusher A. L., Olsen, M. and Nizzetto L. (2018). Validation of a method for extracting microplastics from complex, Organic-Rich, Environmental Matrices. *Environmental Science & Technology*, **52**(13), 7409–7417.

Jemec Kokalj A., Horvat P., Skalar T. and Kržan A. (2018). Plastic bag and facial cleanser derived microplastic do not affect feeding behaviour and energy reserves of terrestrial isopods. *Science of the Total Environment*, **615**, 761–766.

Kacprzak M., Neczaj E., Fijałkowski K., Grobelak A., Grosser A., Worwag M., Rorat A., Brattebo H., Almås Å. and Singh B. R. (2017). Sewage sludge disposal strategies for sustainable development. *Environmental Research*, **156**, 39–46.

Koelmans A. A., Bakir A., Burton G. A. and Janssen C. R. (2016). Microplastic as a vector for chemicals in the aquatic environment: critical review and model-supported reinterpretation of empirical studies. *Environmental Science & Technology*, **50**, 3315–3326.

Lares M., Ncibi M. C., Sillanpää M. and Sillanpää M. (2018). Occurrence, identification and removal of microplastic particles and fibers in conventional activated sludge process and advanced MBR technology. *Water Research*, **133**, 236–246.

Lee H. and Kim Y. (2018). Treatment characteristics of microplastics at biological sewage treatment facilities in Korea. *Marine Pollution Bulletin*, **137**, 1–8.

Leslie H. A., Brandsma S. H., van Velzen M. J. M. and Vethaak A. D. (2017). Microplastics en route: field measurements in the Dutch river delta and Amsterdam canals, wastewater treatment plants, North Sea sediments and biota. *Environment International*, **101**, 133–142.

Li X., Chen L., Mei Q., Dong B., Dai X., Ding G. and Zeng E.Y. (2018). Microplastics in sewage sludge from the wastewater treatment plants in China. *Water Research*, **142**, 75–85.

Liu X., Yuan W., Di M., Li Z. and Wang J. (2019). Transfer and fate of microplastics during the conventional activated sludge process in one wastewater treatment plant of China. *Chemical Engineering Journal*, **362**, 176–182.

LMD KLD and HOD (2019). Forskrift om gjødselvarer mv. av organisk opphav (Regulations on fertilizer products, etc. of organic origin). Ministry of Agriculture and Food (LMD), Ministry of Climate and Environment (KLD) and Ministry of Health and Care Services (HOD), FOR-2003-07-04-951, last amendment 30 January 2019. See: https://lovdata.no/dokument/SF/forskrift/2003-07-04-951 (accessed 12 June 2019).

Lusher A. L., Hurley R. R., Vogelsang C., Nizzetto L. and Olsen M. (2018). Mapping microplastics in sludge. Report No. M907 for the Norwegian Institute for Water Research (NIVA). NIVA, Oslo.

Magni S., Binelli A., Pittura L., Avio C. G., Della Torre C., Parenti C. C., Gorbi S. and Regoli F. (2019). The fate of microplastics in an Italian Wastewater Treatment Plant. *Science of the Total Environment*, **652**, 602–610.

Magnusson K. and Norén F. (2014). Screening of microplastic particles in and down-stream a wastewater treatment plant. Report for the Swedish Environmental Protection Agency (No. C 55). Swedish Environmental Research Institute (IVL), Stockholm.

Mahon A. M., O'Connell B., Healy M. G., O'Connor I., Officer R., Nash R. and Morrison L. (2017). Microplastics in sewage sludge: effects of treatment. *Environmental Science & Technology*, **51**, 810–818.

Mateo-Sagasta J., Raschid-Sally L. and Thebo A. (2015). Global wastewater and sludge production, treatment and use. In: Wastewater: Economic Asset in an Urbanizing World, P. Drechsel, M. Qadir and D. Nad Wichelns (eds.), Springer, Dordrecht, pp. 15–38.

Michielssen M. R., Michielssen E. R., Ni J. and Duhaime M. B. (2016). Fate of microplastics and other small anthropogenic litter (SAL) in wastewater treatment plants depends on unit processes employed. *Environmental Science: Water Research & Technology*, **2**, 1064–1073.

Milieu Ltd, WRC, RPA (2013). Environmental, economic and social impacts of the use of sewage sludge on land. Final Report – Part I: Overview Report. Milieu Ltd, Brussels, Belgium.

Mintenig S. M., Int-Veen I., Löder M. G. J., Primpke S. and Gerdts G. (2017). Identification of microplastic in effluents of waste water treatment plants using focal plane array-based micro-Fourier-transform infrared imaging. *Water Research*, **108**, 365–372.

Mitrano D. M., Beltzung A., Frehland S., Schmiedgruber M., Cingolani A. and Schmidt F. (2019). Synthesis of metal-doped nanoplastics and their utility to investigate fate and behaviour in complex environmental systems. *Nature nanotechnology*, **14**, 362–368.

Murphy F., Ewins C., Carbonnier F. and Quinn B. (2016). Wastewater Treatment Works (WwTW) as a source of microplastics in the aquatic environment. *Environmental Science & Technology*, **50**, 5800–5808.

Napper I. E. and Thompson R. C. (2016). Release of synthetic microplastic plastic fibres from domestic washing machines: effects of fabric type and washing conditions. *Marine Pollution Bulletin*, **112**(1–2), 39–45.

Nizzetto L., Futter M. and Langaas S. (2016). Are agricultural soils dumps for microplastics of urban origin? *Environmental Science & Technology*, **50**, 10777–10779.

Rillig M. C., Ingraffia R. and Machado de S. A A. (2017). Microplastic incorporation into soil in agroecosystems. *Frontiers in Plant Science*, **8**, 1805.

Rodriguez-Seijo A., Lourenço J., Rocha-Santos T. A. P., da Costa J., Duarte A.C., Vala H. and Pereira R. (2017). Histopathological and molecular effects of microplastics in *Eisenia andrei Bouché*. *Environmental Pollution*, **220**, 495–503.

Schmiedgruber M., Hufenus R. and Mitrano D. M. (2019). Mechanistic understanding of microplastic fiber fate and sampling strategies: synthesis and utility of metal doped polyester fibers. *Water Research*, **155**, 443–430.

Sujathan S., Kniggendorf A.-K., Kumar A., Roth B., Rosenwinkel K.-H. and Nogueira R. (2017). Heat and bleach: a cost-efficient method for extracting microplastics from return activated sludge. *Archives of Environmental Contamination and Toxicology*, **73**, 641–648.

Talvitie J., Mikola A., Setälä O., Heinonen M. and Koistinen A. (2017). How well is microlitter purified from wastewater? – A detailed study on the stepwise removal of microlitter in a tertiary level wastewater treatment plant. *Water Research*, **109**, 164–172.

Tchobanoglous G., Burton F. L. and Stensel H. D. (2003). Wastewater Engineering Treatment and Reuse, 4th Ed. Metcalf & Eddy, Inc., McGraw-Hill, NY, ISBN 0-07-112250-8.

Vollertsen J. and Hansen A. A. (2017). Microplastic in Danish wastewater: Sources, occurrences and fate (*Environmental Project No. 1906*). Danish Environmental Protection Agency, Copenhagen.

Wijesekara H., Bolan N. S., Bradney L., Obadamudalige N., Seshadri B., Kunhikrishnan A., Dharmarajan R., Ok Y. S., Rinklebe J., Kirkham M. B. and Vithanage M. (2018). Trace element dynamics of biosolids-derived microbeads. *Chemosphere*, **199**, 331–339.

Water Pollution Control Federation (WPCF) (1985). Clarifier design, WPCF Manual of Practice FD-10. WPCF, Alexandria, VA.

Xu B., Liu F., Brookes P. C. and Xu J. (2018). Microplastics play a minor role in tetracycline sorption in the presence of dissolved organic matter. *Environmental Pollution*, **240**, 87–94.

Yang G., Zhang G. and Wang H. (2015). Current state of sludge production, management, treatment and disposal in China. *Water Research*, **78**, 60–73.

Yang L., Li K., Cui S., Kang Y., An L. and Lei K. (2019a). Removal of Microplastics in Municipal Sewage from China's Largest Water Reclamation Plant. *Water Research*, **155**, 175–181.

Yang Y., Liu G., Song W., Ye C., Lin H., Li Z. and Liu W. (2019b). Plastics in the marine environment are reservoirs for antibiotic and metal resistance genes. *Environment International*, **123**, 79–86.

Zhu D., Chen Q.-L., An X.-L., Yang X.-R., Christie P., Ke X., Wu L.-H. and Zhu Y.-G. (2018). Exposure of soil collembolans to microplastics perturbs their gut microbiota and alters their isotopic composition. *Soil Biology and Biochemistry*, **116**, 302–310.

Zubris K. A. V. and Richards B. K. (2005). Synthetic fibers as an indicator of land application of sludge. *Environmental Pollution*, **138**, 201–211.

Chapter 7

Modeling microplastics transport and fate in the marine environment around a wastewater effluent discharge pipe

D. P. Korfiatis

University of Patras, Physics Department, Patras, Greece

Keywords: Advection, Aggregation, Degradation, Diffusion, Mathematical model, Sedimentation, Wastewater Treatment Plants (WWTPs)

7.1 INTRODUCTION

Plastics are organic polymers which are made of oil and/or natural gas. Their extensive use and nearly inexhaustible applications are well known. The world production of plastic materials has been constantly increasing during the last decades (Horton & Dixon, 2018).

The term "microplastics" refers to small fragments of plastic debris. Today, microplastics have accumulated in the environment on a global scale. Pollution of the marine environment due to microplastics is the focus of scientific research because of its great ecological importance (Frere *et al.*, 2017). The accumulation of microplastics in the marine environment, quite apart from the negative aesthetic effect, also has both direct and indirect harmful effects on ocean biosystems (Auta *et al.*, 2017; Ogunola & Palanisami, 2016).

There is no agreement among researchers about the size definition of microplastics and nanoplastics. In this study, microplastics are defined as plastic particles lower than 5 mm in any one dimension, and nanoplastics as plastic particles in the size range of 1–100 nm (GESAMP, 2015; Rios Mendoza *et al.*, 2018). Plastics manufactured in miniature size are called primary microplastics.

Editors: Hrissi K. Karapanagioti and Ioannis K. Kalavrouziotis
doi: 10.2166/9781789061697_0101

However, the majority of microplastics in the marine environment are secondary microplastics which result from the fragmentation of mesoplastic or larger plastic waste (Cole *et al.*, 2011). This breakdown of larger plastic items can occur through a variety of mechanisms, and the degradation processes can be classified as mechanical, chemical and biological (Andrady, 2011; Horton & Dixon, 2018).

Rivers have been thoroughly investigated as microplastic marine litter sources in several research papers (Besseling *et al.*, 2017; Schmidt *et al.*, 2017; Siegfried *et al.*, 2017). In a recent study, Mourgkogiannis *et al.* (2018) showed that the quantity of microplastics ending up in sea environments due to wastewater treatment plants is quite significant. Thus, wastewater treatment plants must also be considered as microplastic marine litter sources. The ways in which microplastics are introduced into wastewater treatment plants have been reported by Karapanagioti (2017).

It is widely accepted that numerical modeling of the fate and transport of microplastics in the ocean environment has a crucial role in the understanding of the problem and in finding possible solutions (Hardesty *et al.*, 2017). Mathematical models developed for freshwater environments (Besseling *et al.*, 2017) are based on the shallow water St. Venant equations. They also include the transformation processes of microplastics and of sedimentation.

In this chapter, the mathematical modeling of the transformation and transport processes of microplastics in the marine environment around a wastewater effluent discharge pipe is studied.

7.2 TRANSFORMATION PROCESSES

7.2.1 Homoaggregation

The aggregation of similar particles with each other is called homoaggregation. According to the model proposed by Smoluchowski (1917), the rate of homoaggregation of plastic particles in the sea environment can be calculated by Equation (7.1):

$$\frac{dn_j}{dt} = \frac{1}{2}\sum_{i=1}^{j-1} a_{i,j-i} \cdot K_{i,j-i} \cdot n_i \cdot n_{j-i} - n_j \sum_{i=1}^{\infty} a_{i,j} \cdot K_{i,j} \cdot n_i \tag{7.1}$$

where: n_j is the number of particles of size class j, in giga particles per m^3 (10^9 particles m^{-3}), $a_{i,j}$ is the attachment efficiency of particles of size class i with particles of size class j, and $K_{i,j}$ is the collision frequency of particles of size class i with particles of size class j, in m^3 giga particles^{-1} s^{-1}.

The first term on the right-hand side of the equation describes the formation of particles of size class j through agglomeration of particles i and $j - i$. The second term describes the loss of particles of size class j through agglomeration with particles i of other size classes to form larger particles.

The collision rate is determined mainly by the following three processes: Brownian motion (peri-kinetic); fluid motion (ortho-kinetic); and differential settling (Quik *et al.*, 2014). The collision frequency, $K_{i,j}$, is given by Equation (7.2):

$$K_{i,j} = \left(\frac{2k_B T}{3\mu} \frac{(a_i + a_j)^2}{a_i a_j} + \frac{4}{3} G(a_i + a_j)^3 + \left(\frac{2\pi g}{9\mu} \right) (\rho_p - \rho_w)(a_i + a_j)^3 (a_i - a_j) \right) \cdot 10^9 \tag{7.2}$$

where: k_B is Boltzmann's constant, T is the absolute temperature, μ is the dynamic viscosity coefficient of the water, α is the radius of the particle, G is the shear rate, g is the gravitational acceleration, and ρ_p and ρ_w are the particle density and water density, respectively.

Although homoaggregation is a very important transformation process for microplastics in many situations, this is not the case for wastewater treatment plants. The concentrations of microplastics ending up in the marine environment from a wastewater effluent discharge pipe are very much lower than from other plastic marine litter sources. Worst-case values for the concentration of microplastics flowing from a wastewater treatment plant to the sea are between 100 and 50,000 particles per m^3 (Mourgkogiannis *et al.*, 2018).

From Smoluchowski (see Equation 7.1 above), it is deduced that for very low microplastic particle concentrations the probability for homoaggregation is quite low. Besseling *et al.* (2017) arrived at the same conclusion through numerical calculations. So, in a first approach, it is valid to ignore the homoaggregation process.

7.2.2 Heteroaggregation

Heteroaggregation is the process by which microplastics form aggregates with suspended solids (SS) in wastewater effluent. The mathematical description of heteroaggregation is similar to that of homoaggregation (Besseling *et al.*, 2017; Quik *et al.*, 2014). So, the rate of change of density of plastic particles of size class j due to heteroaggregation is given by Equation (7.3):

$$\frac{dn_j}{dt} = -a_{het} n_j \sum_{i=1}^{m} K_{j,SSi} n_{SSi} \tag{7.3}$$

where m is the number of size classes of suspended solids considered in the model, and n_{SSi} the particle density of suspended solids of the i size class which have a radius of a_{SSi}.

The collision frequency, $K_{j,SSi}$, is given by:

$$K_{j,SSi} = \left(\frac{2k_B T}{3\mu} \frac{(a_j + a_{SSi})^2}{a_j a_{SSi}} + \frac{4}{3} G(a_j + a_{SSi})^3 + \pi (a_j + a_{SSi})^2 |v_{s,j} - v_{s,SSi}| \right) \cdot 10^9 \tag{7.4}$$

The shear rate (G) which appears in Equation (7.4) implies laminar flow. Shear rate is strongly dependent on weather conditions and sea water motion (Arvidsson *et al.*, 2011). Sedimentation rates are denoted by $v_{s,j}$ and $v_{s,SSi}$ for microplastics and suspended solids, respectively. The collision frequencies of microplastic particles with suspended solids are higher in the sediment compared to in the water phase (Besseling *et al.*, 2017).

7.2.3 Degradation

In first order, the degradation of microplastics can be modeled through a degradation rate constant, $k_{\deg j}$ (Besseling *et al.*, 2017), as:

$$\frac{dn_j}{dt} = -k_{\deg j} n_j \tag{7.5}$$

Thus, the degradation rate of microplastics can be assumed to be proportional to the existing particle concentration.

7.3 TRANSPORT

7.3.1 Sedimentation

The rate of sedimentation of microplastics of each size class is analogous to the particle concentration of that size class according to the relation given in Equation (7.6):

$$\frac{dn_j}{dt} = -\frac{v_{s,j}}{d_j} n_j \tag{7.6}$$

where: d_j is the sedimentation length and v_{sj} is the sedimentation velocity of the particles of size class j. Sedimentation velocity can be calculated through Stokes' law (Equation 7.7):

$$v_{s,j} = \frac{2a_j^2(\rho_p - \rho_w)g}{9\mu} \tag{7.7}$$

Certainly, the sedimentation rate is dependent on the size of the particles. As the particle size increases, so does the sedimentation rate. So, a correction factor determined by the particle diameter can therefore be added to Equation (7.6) (Arvidsson *et al.*, 2011).

7.3.2 Advection–diffusion

The advection–diffusion of the microplastics in a one-dimensional model can be described by the partial differential equation expressed by Equation (7.8):

$$\frac{\partial n}{\partial t} + v\frac{\partial n}{\partial x} = D\frac{\partial^2 n}{\partial x^2} - Kn \tag{7.8}$$

where Kn represents the removal rate of microplastics from the water phase. The homoaggregation process is ignored because of the low microplastic concentration, as justified above. Also, only one size class of particles is considered, for clarity. Under these assumptions, it is obvious from the equations describing the transformation processes and sedimentation that the total rate of loss of particles is analogous to particle density. The overall proportion factor is K. Certainly K is dependent on the concentrations of suspended solids particles. It is also dependent on the concentrations of microplastic particles of other size classes if the homoaggregation process is taken into account.

The diffusion coefficient of microplastics in the marine environment is D and v is the velocity of microplastic particles at a position x (distance from the outlet pipe). For simplicity, only one-dimensional motion has been considered, ignoring lateral diffusion.

Under steady state conditions, Equation (7.8) is converted to an ordinary differential equation as given in Equation (7.9):

$$D\frac{d^2n}{dx^2} - v\frac{dn}{dx} - Kn = 0 \tag{7.9}$$

and $n(x=0)=n_0$ (the microplastics concentration at the pipe outlet) and $n=0$ as $x\rightarrow\infty$ are specified as boundary conditions.

Assuming a constant velocity, v, and also a constant factor, K, Equation (7.9) leads to a simple exponential decrease of the microplastics concentration with x, as expressed by Equation (7.10):

$$n = n_0 \cdot e^{-\lambda x} \tag{7.10}$$

where:

$$\lambda = \frac{\sqrt{v^2 + 4DK} - v}{2D} \tag{7.11}$$

In a more realistic model, the reduction of velocity as a function of position x is taken into account. As microplastics move in the marine environment, their motion is slowed down due to the opposite drag force from the water. The motion is governed through Newton's Second Law:

$$m\frac{dv}{dt} = -F_D \tag{7.12}$$

Assuming laminar flow (low values of Reynolds number) the drag force is given by:

$$F_D = \frac{6A\mu}{a}v \tag{7.13}$$

Solving Equation (7.12) and after elimination of time, t, between v and x we get a linearly reducing velocity with displacement x of the form:

$$v = v_0 - \frac{6A\mu}{ma}x \tag{7.14}$$

where v_0 is the outflow velocity of microplastics from the pipe (at $x = 0$).

At sufficiently high values of Reynolds number, the drag force is approximately proportional to v^2. This flow regime is sometimes referred to as the Newton's Law region. The drag force takes the form:

$$F_D = C \cdot v^2 \tag{7.15}$$

where C is a constant dependent on sea water density and the particle's cross-sectional area.

In this case, the dependence of velocity on x can be proved in the form:

$$v = v_0 \cdot e^{-\frac{C}{m}x} \tag{7.16}$$

In other words, velocity decreases exponentially with x. Replacing v in Equation (7.9) with Equation (7.14) or Equation (7.16) (depending on the value of Reynolds number) could lead to a more accurate calculation of the distribution of microplastics concentration along the x direction.

7.4 MATHEMATICAL MODEL

In order to describe the fate and transport of microplastics which outflow to the sea environment through a wastewater effluent discharge pipe, the advection–diffusion Equation (7.9) has to be used for a number of particle size classes. We are interested in steady state under permanent flow conditions and spherical particles are assumed. In this way, the result is a set of ordinary differential equations:

$$D_j \frac{d^2 n_j}{dx^2} - v_j(x)\frac{dn_j}{dx} - K_j n_j = 0 \tag{7.17}$$

where $j = 1, \ldots, l$

and

$$K_j = a_{het} \sum_{i=1}^{m} K_{j,SSi} n_{SSi} + k_{\deg j} + \frac{v_{sj}}{d_j} \tag{7.18}$$

Not taking into account homoaggregation has led to a system of uncoupled equations. The symbol l denotes the number of the size classes used in the model and its value is case dependent. For the determination of the appropriate value of n in a specific case, data about the microplastic concentration of the wastewater effluent discharge pipe under consideration are required.

Solution of the system gives the function $n_j(x)$ representing the distribution of microplastics of any of the l size classes at a distance, x, from the end of the pipe. The boundary conditions are: $n_j(x=0)=n_{0j}$ the microplastics concentration of size j at the pipe outlet and $n_j=0$ as $x\rightarrow\infty$, $j=1, \ldots l$.

The velocity $v_j(x)$ as a function of the distance x can be inserted into Equation (7.17) using Equation (7.14) or Equation (7.16) and is dependent on the particle size class. The initial velocity at the end of the pipe, which is independent of the particle size, can be simply calculated from the relation:

$$v_0 = \frac{Q}{A} \tag{7.19}$$

where Q is the outlet volumetric flow rate and A is the pipe's cross-sectional area. The values of Q along with n_{oj} can be provided from the specific wastewater treatment plants under consideration.

The solution of the system of equations is the distribution of microplastics concentration along the x-direction. From that distribution, the maximum distance from the pipe outlet in the sea where microplastics have a detectable concentration can easily be deduced.

It should be noted that, for a more sophisticated model of a specific wastewater treatment plant, the position of the pipe exit with respect to wind direction has to be taken into account (Critchell & Lambrechts, 2016).

7.5 CONCLUSIONS

This chapter aimed to study the fate and transport of microplastics which escape from wastewater treatment plants to marine environment and presented a mathematical model to describe the dispersion of microplastics in the sea. Solution of the system of differential equations gives information about the distance from the end of the discharge pipe at which microplastics concentration is considerable.

The model presented is based on the advection–diffusion equation in which the opposite drag force from water has been incorporated, and the transformation processes of microplastics and sedimentation are also included.

REFERENCES

Andrady A. L. (2011). Microplastics in the marine environment. *Marine Pollution Bulletin* **62**, 1596–1605.

Arvidsson R., Molander S., Sandén B. A. and Hassellöv M. (2011). Challenges in exposure modeling of nanoparticles in aquatic environments. *Human and Ecological Risk Assessment* **17**, 245–262.

Auta H. S., Emenike C. U. and Fauziah S. H. (2017). Distribution and importance of microplastics in themarine environment: a review of the sources, fate, effects, and potential solutions. *Environment International* **102**, 165–176.

Besseling E., Quik J. T. K., Sun M. and Koelmans A. A. (2017). Fate of nano- and microplastic in freshwater systems: a modeling study. *Environment Pollution* **220**, 540–548.

Cole M., Lindeque P., Halsband C. and Galloway T. S. (2011). Microplastics as contaminants in the marine environment: a review. *Marine Pollution Bulletin* **62**, 2588–2597.

Critchell K. and Lambrechts J. (2016). Modelling accumulation of marine plastics in the coastal zone; what are the dominant physical processes? *Estuarine Coastal and Shelf Science* **171**, 111–122.

Frere L., Paul-Pont I., Rinnert E., Petton S., Jaffre J., Bihannic I., Soudant P., Lambert C. and Huvet A. (2017). Influence of environmental and anthropogenic factors on the composition, concentration and spatial distribution of microplastics: a case study of the Bay of Brest (Brittany, France). *Environment Pollution* **225**, 211–222.

GESAMP (2015). Sources, fate and effects of microplastics in the marine environment: a global assessment. In: (IMO/FAO/UNESCO–IOC/UNIDO/WMO/IAEA/UN/UNEP/UNDP Joint Group of Experts on the Scientific Aspects of Marine Environmental Protection (GESAMP)), P. J. Kershaw, (ed.), *Rep. Stud. GESAMP No. 90*, 96p.

Hardesty B. D., Joseph Harari J., Atsuhiko Isobe A., Laurent Lebreton L., Maximenko N., Potemra J., van Sebille E., Vethaak A. D. and Wilcox C. (2017). Using Numerical Model Simulations to Improve the Understanding of Micro-plastic Distribution and Pathways in the Marine Environment. *Frontiers in Marine Science* **4**, 30.

Horton A. A. and Dixon S. J. (2018). Microplastics: an introduction to environmental transport processes. *WIREs Water* **5**, e1268.

Karapanagioti H. K. (2017). Microplastics and synthetic fibers in treated wastewater and sludge. In: Wastewater and Biosolids Management, I. K. Kalavrouziotis, (ed.), IWA Publishing, London, pp. 77–88.

Mourgkogiannis N., Kalavrouziotis I. K. and Karapanagioti H. K. (2018). Questionnaire-based survey to managers of 101 wastewater treatment plants in Greece confirms their potential as plastic marine litter sources. *Marine Pollution Bulletin* **133**, 822–827.

Ogunola O. S. and Palanisami T. (2016). Microplastics in the Marine Environment: current Status, Assessment Methodologies, Impacts and Solutions. *Journal of Pollution Effects Control* **4**, 161.

Quik J. T. K., Van de Meent D. and Koelmans A. A. (2014). Simplifying modeling of nanoparticle aggregation-sedimentation behavior in environmental systems: a theoretical analysis. *Water Research* **62**, 193–201.

Rios Mendoza L. M., Karapanagioti H. and Ramirez Alvarez N. (2018). Micro(nanoplastics) in the marine environment: current knowledge and gaps. *Current Opinion in Environmental Science & Health* **1**, 47–51.

Schmidt C., Krauth T. and Wagner S. (2017). Export of plastic debris by rivers into the sea. *Environmental Science & Technology* **51**, 12246–12253.

Siegfried M., Koelmans A. A., Besseling E. and Kroeze C. (2017). Export of microplastics from land to sea. A modelling approach. *Water Research* **127**, 249–257.

Smoluchowski M. (1917). Versuch einer matematischen Theorie der Koagulationskinetic kolloider Losungen (Attempt of a mathematical theory of coagulation kinetic colloidal solutions). *Z Phys Chem* **92**, 129–168.

Chapter 8

Evaluating wastewater effluent as a source of microplastics in environmental samples

W. Cowger[1], A. B. Gray[1], M. Eriksen[2], C. Moore[3] and M. Thiel[4]

[1]Riverside, Environmental Science, University of California, Riverside, USA
[2]5 Gyres Institute, Los Angeles, USA
[3]Algalita, Long Beach, USA
[4]Department of Marine Biology, Universidad Catolica del Norte, Coquimbo, Chile

Keywords: Anthropogenic litter, Marine debris, Mismanaged waste, Plastic pollution, Source allocation, Wastewater

8.1 INTRODUCTION

Microplastic's (GESAMP, 2015) association with wastewater discharge predates 21st Century concerns with microplastic pollution. Synthetic clothing fibers and plastic microbeads from household laundering and consumer products have long been routed to wastewater treatment plants where it is reported that most are captured in the sewage sludge and the remainder emitted in the effluent (Fendal & Swell, 2009; Gregory, 1996; Habib *et al.*, 1998; Ziajahromi *et al.*, 2016). In the late 1990s synthetic fibers from clothing were proposed as an indicator of wastewater environmental fluxes after it was found that concentrations of synthetic fibers decreased with increasing distance from sewage sludge application or from outfalls (Habib *et al.*, 1998). Today synthetic fibers are a recognized form of pollution that are increasingly monitored by environmental

doi: 10.2166/9781789061697_0109

scientists (Browne *et al.*, 2011; Miller *et al.*, 2017), and microbeads have been banned from use in personal care products (CA State Legislature, 2015; US Congress, 2015). Over the ensuing decades we have learned that synthetic fibers also originate from atmospheric deposition (Baldwin *et al.*, 2016; Dris *et al.*, 2016). With this discovery, and further elucidation of other sources of plastic pollution, the ability to identify microplastic sources from environmental samples has come under increasing scrutiny (Leslie *et al.*, 2017).

Research papers assessing the provenance of microplastics in environmental samples have reported a wide range of confidence in their ability to attribute sources. Some reports have stated that locating the source of plastics from environmental samples is impossible, or that there is a high level of uncertainty in the assessment (Claessens *et al.*, 2011; Leslie *et al.*, 2017; Woodall *et al.*, 2014). Other reports state with confidence that the microplastics they observed in the environment originated from wastewater effluent (Estahbanati & Fahrenfeld, 2016; Vermaire *et al.*, 2017; Warrack *et al.*, 2018), or not (Campbell *et al.*, 2017). The differences in true levels of certainty arise primarily from differences in the technique used to identify sources.

To address the contribution of study design to this microplastic source uncertainty, 23 papers assessing wastewater effluent as a source of microplastics to the environment have been reviewed using evidence gathered from samples collected outside of the effluent discharge point. For details on the physical techniques for microplastics sampling strategies, see: Hidalgo-Ruz *et al.* (2012); Blair *et al.* (2017); Li *et al.* (2017); Hanvey *et al.* (2017); Shim *et al.* (2017); Mai *et al.* (2018); and Silva *et al.* (2018). This chapter focuses on the following question: *Is the way we are ascribing source to wastewater using environmental samples accurate, and how can it be improved?* In this context, source can be described as an absolute source (with an exact number of microplastics coming from wastewater) or as a relative source (with a proportion of microplastics from wastewater compared to from another source). The papers reviewed all assessed relative sources. As in other areas of interest in microplastics research, the number of papers that have discussed wastewater effluent source allocation has risen in the past few years (Gago *et al.*, 2018) (Figure 8.1). The 23 papers reviewed here represent globally distributed regions (Figure 8.1) and all continents are represented except Oceania and Antarctica. Eighteen of the papers suggested that wastewater treatment plant effluent is a source of microplastics in their study region (Table 8.1). Eleven of the papers explicitly stated that determining the source of plastic was a primary goal (Table 8.1). The studies leveraged samples from surface water, sediment, the water column and organisms to assess plastic pollution in streams, coastal oceans, lakes and estuaries. Marine environments were studied in eight of the reviewed papers, whilst the remainder studied freshwater. Six of the studies compared environmental samples to wastewater effluent samples (Table 8.1). In this critical review, the techniques used to assess sources of microplastics in the environment were reviewed and provide a framework for how

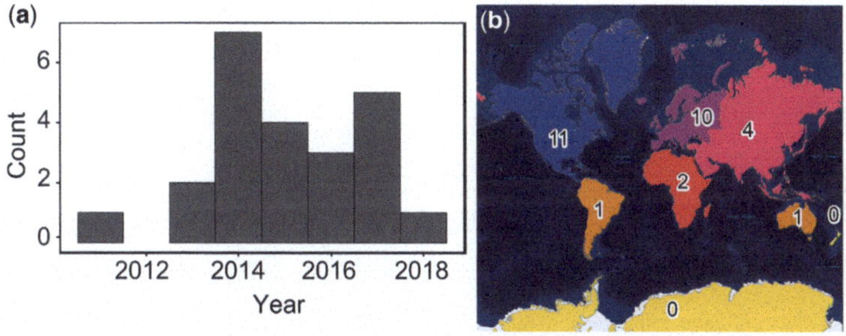

Figure 8.1 (a) Histogram of annual report numbers from the 23 studies in this review. (b) number of papers in each continent from this review.

these claims can be weighed by the scientific community, from lowest certainty (anecdotal evidence) to highest (full mass balance evaluation) (Figure 8.2).

8.2 ANECDOTAL EVIDENCE

Anecdotal evidence is obtained from experience, unverified third-party reports or *ad hoc* estimation of potential sources and is the point where most scientific inquiry begins. Although most scientists would agree that anecdotal evidence should not be a primary basis of scientific conclusions, sometimes this is the case. Free *et al.* (2014) and Zhang *et al.* (2015) collected plastics in populated regions that had no wastewater treatment plants and determined that, by default, wastewater could not be a source of the plastic pollution they found. Lechner *et al.* (2014) attributed the source of the plastic pollution they found to wastewater effluents because their sampling locations were downstream of wastewater treatment plants that were of low purification (primary and secondary treatment). These are contrasting examples of using anecdotal evidence. While Zhang *et al.* (2015) and Free *et al.* (2014) determined that a lack of municipal wastewater treatment and poor wastewater quality indicated a low likelihood for the source of the plastics to be from wastewater, Lechner *et al.* (2014) reached the opposite conclusion. This contrast seems to be centered around the question, "What is wastewater?"

We advance the argument that wastewater effluent should be defined as any sewage water discharged by humans and agree with Lechner *et al.* (2014) that a lack of treatment or lower quality treatment corresponds to a higher likelihood for wastewater discharge contaminated by plastics. In future studies, the degree of wastewater treatment could be classified as informal or formal (with an in-depth explanation of the degree of treatment) and evaluated in terms of connectivity to aquatic systems. For example, in developing countries with no formal treatment, waste effluent with high connectivity to receiving water bodies such as open sewers and clothes washing in streams could represent a substantial source of informal wastewater-based microplastics. Additionally, an

Table 8.1 Metadata from reviewed publications.

Source	Paper Type	Publisher Name	Quantification Technique	Taxonomy Techniques	Size Range	Density Separation	OM Destruction	Field Methods	Smallest Sample Volume (L)
Baldwin *et al.* (2016)	Article	Environmental Science and Technology	Count	Shape (Fibers, Fragments, Pellets/Beads, Foam, Film), Size (0.355–0.99, 1–4.75, >4.75)	>333 μm	NA	H_2O_2	Net	1000+
Browne *et al.* (2011)	Article	Environmental Science and Technology	Count	Polymer (Polyester, Acrylic, Polypropylene, Polyethylene, Polyamide)	NA	NaCl	NA	Grab Sample	0.25
Campbell *et al.* (2017)	Note	FACETS	Count	Shape (Fibers, Fragments, Beads, Foam, Film)	>5 μm and >80 μm water	NA	Yes	Traps	500
Castañeda *et al.* (2014)	Rapid Communication	NRC Press	Count	Shape (Microbeads)	>500 μm	NA	NA	Grab Sample	2.25
Claessens *et al.* (2011)	Article	Marine Pollution Bulletin	Count	Shape (Fibers, Granules, Films, Spherules)	>38 μm – 1 mm	NaCl	NA	Cores	0.5
Dris *et al.* (2015)	Research Paper	CSIRO Publishing	Count	Size (100–500, 500–1,000, 1,000–5,000)	>80 μm and >330 μm	NA	NA	Net	450
Dubaish and Liebzet (2013)	Research Article	Water Soil and Air Pollution	Count	Shape (Granular, Fibers)	>40 μm	NA	H_2O_2	Grab Sample	0.1
Eriksen *et al.* (2013)	Article	Marine Pollution Bulletin	Count	Shape (Fragment, Film, Foam, Pellet, Line), Size (0.355–0.999, 1–4.749, >4.75)	>333 μm	NA	NA	Net	1000+
Estahbanati and Fahrenfeld (2016)	Article	Chemosphere	Count	Shape (Primary, Secondary), Size (63–125, 125–250, 250–500, 500–2,000)	>125 μm	NaCl	H_2O_2	Net	1000+
Free *et al.* (2014)	Article	Marine Pollution Bulletin	Mass and Count	Shape (Fragment, Line/Fiber, Pellet, Film, Foam), Size (0.355–0.999, 1–4.749, >4.75)	>333 μm	NaCl	H_2O_2	Net	1000+
Gallagher *et al.* (2016)	Article	Marine Pollution Bulletin	Count	Shape (Fibre, Rounded, Irregular, Oval), Color (Black, White, Clear, Blue, White/Blue, Grey, Yellow, Green, Orange, Brown, Blue/Black)	>300 μm	NaCl	NA	Net	1000+
Lechner *et al.* (2014)	Short Communication	Environmental Pollution	Mass and Count	Shape (Pellets, Flakes, Spherules, Other)	>500 μm	Yes	NA	Net	1000+

Source	Access	Goal Includes Effluent or Source Allocation	Is Effluent a Contributor?	Technique	Media	Ecosystem	Country	Chemical Confirmation	Sampled Wastewater
Baldwin *et al.* (2016)	Yes	Yes	No	Correlation	Surface water	Stream	USA	No	No
Browne *et al.* (2011)	No	Yes	Yes	Taxonomic	Sediment	Shorelines	Multiple	FTIR	Yes
Campbell *et al.* (2017)	Yes	Yes	No	Correlation	Organism and Water	Stream	Canada	No	No
Castañeda *et al.* (2014)	No	No	Yes	Taxonomic	Sediment	Stream	Canada	No	No
Claessens *et al.* (2011)	Yes	No	Yes	Taxonomic	Sediment	Coastal Ocean	Belgium	FTIR	No
Dris *et al.* (2015)	No	Yes	No	Taxonomic	Surface water and Atmosphere	River	France	No	Yes
Dubaish and Liebezeit (2013)	No	No	Yes	Taxonomic	Surface Water	Ocean	North Sea	No	No
Eriksen *et al.* (2013)	No	No	Yes	Taxonomic	Surface water	Lake	USA	No	No
Estahbanati *et al.* (2016)	No	Yes	Yes	Correlation	Surface water	Stream	USA	No	No
Free *et al.* (2014)	Yes	Yes	No	Anecdote and Taxonomic	Surface water	Lake	Mongolia	No	No
Gallagher *et al.* (2016)	No	No	Yes	Taxonomic	Water column	Estuary	UK	FTIR	No
Lechner *et al.* (2014)	No	No	Yes	Anecdote	Surface water	Stream	Europe	No	No

(Continued)

Table 8.1 Metadata from reviewed publications. (*Continued*)

Source	Paper Type	Publisher Name	Quantification Technique	Taxonomy Techniques	Size Range	Density Separation	OM Destruction	Field Methods	Smallest Sample Volume (L)
Leslie *et al.* (2017)	Research Article	Environment International	Count	Shape (Fibres, Spheres and Foils), Size (10–300, >300–5,000)	10–5,000 µm	NaCl	NA	Continuous Centrifuge	NA
Magnusson and Noren (2014)	Report	Swedish Environmental Research Institute	Count	Shape (Fibers, Particles, Flakes)	>300 um	NA	NA	Net	6600
McCormick *et al.* (2014)	Article	Environmental Science and Technology	Count	Bulk	>333 µm	NaCl	H_2O_2	Net	1000+
Miller *et al.* (2017)	Research Article	Marine Pollution Bulletin	Count	Color (Blue, Red, Black, Transparent, Other), Shape (Fiber, Round, Other), Length (100–1.5 mm, 1.6–3.2 mm, 3.3–9.6 mm)	> 100 um	NA	NA	Grab Sample	1
Smith *et al.* (2017)	Report	Mohawk Watershed Symposium	Count	Shape (Fibers, Films, Pellets, Foams, Fragments)	>333 µm	NaCl	H_2O_2	Net and Grab Sample	1000+
Talvitie *et al.* (2015)	Article	Water Science and Technology	Count	Shape (Fibers, Particles)	>20 um and >200 um	NaCl	NA	Filtration	20
Vermaire *et al.* (2017)	Article	FACETS	Count	Shape (Fibers, Microbeads, Other Plastics)	>100 µm	NaCl	H_2O_2	Net, Grab Sample, Pump	3.5
Warrack *et al.* (2018)	Research Article	Frontiers of Undergraduate Research	Count	Shape (Fragments, Foams, Fibers, Pellets, Films)	>333 µm	NA	H_2O_2	Net	1000+
Woodall *et al.* (2014)	Research Article	Royal Society Open Science	Count	Polymer (Polyester, Acrylic, Other Synthetics), Shape (Fiber), Color (All)	>32 µm	NaCl	NA	Cores	0.01
Zhang *et al.* (2015)	Research Article	Environmental Pollution	Count	Shape (Fragment, Sheet, Line, Foam, Microbeads), Polymer (PE, PP, PS), Size (112–300, 400–500, 500–1,600, 1,600–5,000)	>112 µm	Yes	NA	Net	2000
Zhao *et al.* (2014)	Baseline	Marine Pollution Bulletin	Count	Shape (Fibres, Films, Granules, Spherules), Size (>0.5–1 mm, 1–2.5, 2.5–5, >5)	>333 µm	ZnCl	H_2O_2	Net and Filtration	20

Source	Access	Goal Includes Effluent or Source Allocation	Is Effluent a Contributor?	Technique	Media	Ecosystem	Country	Chemical Confirmation	Sampled Wastewater
Leslie *et al.* (2017)	Yes	Yes	Yes	Taxonomic	Many	Ocean and Stream	Netherlands	FTIR	Yes
Magnusson and Noren (2014)	No	Yes	Yes	Correlation	Surface water	Stream	Sweden	FTIR	Yes
McCormick *et al.* (2014)	No	Yes	Yes	Correlation and Other Sources	Surface water	Stream	USA	No	No
Miller *et al.* (2017)	No	No	Yes	Taxonomic and Correlation	Surface water	Stream	USA	FTIR	No
Smith *et al.* (2017)	No	No	Yes	Taxonomic and Correlation	Surface water and Sediment	Stream	USA	Raman	No
Talvitie *et al.* (2015)	No	Yes	Yes	Other Sources	Water column and Sediment	Coastal Ocean	Finland	No	Yes
Vermaire *et al.* (2017)	No	Yes	Yes	Correlation	Sediment and Water	Stream	Canada	No	Yes
Warrack *et al.* (2013)	No	No	Yes	Correlation	Surface water	Stream	Canada	No	No
Woodall *et al.* (2014)	Yes	No	Yes	Taxonomic	Sediment	Ocean	Multiple	FTIR	No
Zhang *et al.* (2015)	No	No	No	Anecdote and Taxonomic	Surface water	Stream	China	FTIR	No
Zhao *et al.* (2014)	No	No	Yes	Taxonomic	Surface water	Estuary and Coastal Ocean	China	No	No

Figure 8.2 From left to right: this chapter's section headings, which correspond to a spectrum of increasing certainty from source evaluation techniques.

increasing degree of formal wastewater treatment has been found to decrease the concentration of microplastics discharged by effluent (Carr *et al.*, 2016). However, there is a large range of variability in wastewater purification techniques, particularly tertiary treatment, which can be very effective when microfiltration technologies are employed. Thus, further information on whether and by what processes wastewater is treated is of great importance for understanding wastewater derived microplastic discharges. The contribution of wastewater-borne microplastics also depends upon the connectivity between discharge and receiving body. If plastic transport time/distance from the site of waste generation to the sampled receiving body is long relative to the characteristic transport length of microplastic particles (Pizzuto *et al.*, 2017), small communities with a lower level of development (such as those studied by Free *et al.* (2014) and Zhang *et al.* (2015)) may indeed contribute a lower flux of wastewater-borne microplastics to a given waterbody relative to a plumbed system conveying primary waste.

These issues highlight the fact that anecdotal evidence is not the most effective or accurate means of determining the source of microplastics and should be used with caution. More accurate and quantitative assessments of microplastic sources and their connectivity to the environmental system of interest can often be employed.

8.3 TAXONOMIC EVIDENCE

Taxonomic approaches utilize the characteristics of microplastics in the environment, such as the shapes of microbeads and fibers (Figure 8.3), to assess microplastic provenance. Depending on the specific approach, taxonomic evidence can provide qualitative to semi-quantitative evidence of source. Of the 23 papers reviewed, 14 used some type of taxonomic evidence to assign source. Taxonomic groups from microplastics include shape, size, color, polymer type and item type. Unfortunately, the nomenclature used for these taxonomies are not standardized, and in cases where taxonomies are standard their ability to be used to determine source has come into question (Leslie *et al.*, 2017). However, by utilizing multiple forms of taxonomic evidence from microplastics, macroplastics (plastics >5 mm) and nonplastics, confidence provided by taxonomic evidence may be increased.

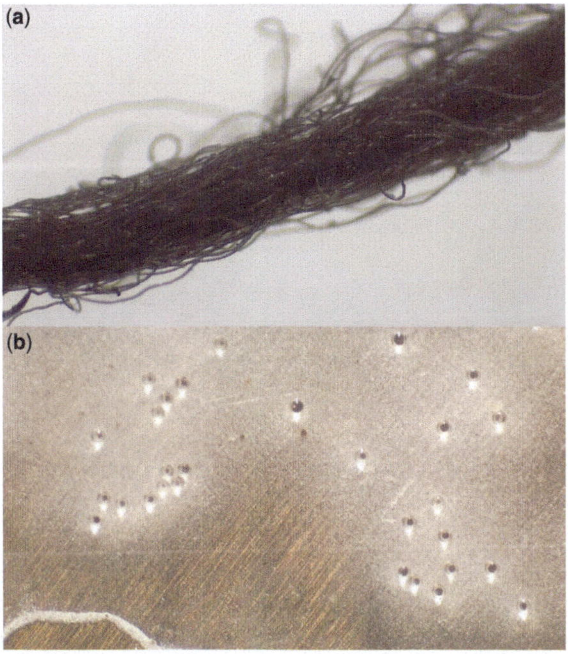

Figure 8.3 Examples of taxonomic classifications. (a) microfibers from clothing; (b) 90 μm clear plastic microbeads surrounded by 2 μm microbeads creating a halo glow (Photo credit: Win Cowger).

8.3.1 Microplastic indicators

A lack of standardized taxonomy has resulted in 19 unique categorization terms used in the literature (Figure 8.4). Though many studies do not explicitly define their taxonomic classifications, we used our best judgement and context or image examples to define the relationships between them in Figure 8.4. Ambiguity introduced from the choices of taxa and their definitions can impede cross-study comparisons and larger scale meta-analysis of existing microplastic datasets. The utility and comparability of taxonomic features would benefit from the optimization of an effective, standardized taxonomic scheme (Helm, 2017).

Some plastic taxonomies overlap in their definition. "Microbeads" and "pellets" seem to overlap in their spherical shape definition but differ based on size. Castañeda *et al.* (2014) only quantified microbeads and included particles up to 2 mm in diameter. Eriksen *et al.* (2013) analyzed consumer microbeads and classified any spherical particles in their environmental samples smaller than 1 mm as microbeads. Fendall and Sewell (2009) found very few cosmetic microbeads larger than 1 mm. We suggest that the spherical plastic particles size threshold between microbeads and pellets should be 1 mm which is a common

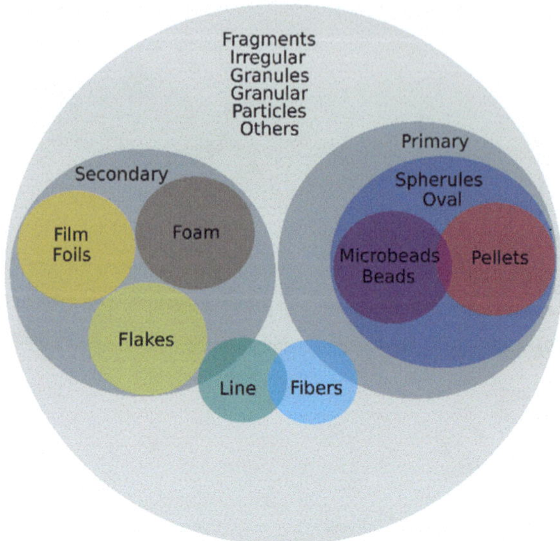

Figure 8.4 Relationships between taxonomic nomenclature. Overlapping circles represent shared definitions within studied publications. Multiple words in a circle means that those words have been used to represent the same classification.

boundary used to distinguish microplastics from mesoplastics (1–25 mm), and that all studies should report the size thresholds used to make this taxonomic distinction. Similarly, there is little that distinguishes "fibers" from "line" and some papers use the two interchangeably (Free *et al.*, 2014). Fibers and line classifications could be separated by defining fibers as from a cloth origin and line as a linear fragment, and by creating a new category for monofilament fishing line. A rigorous study on the uncertainties involved in microplastic taxonomies would greatly benefit the field (Helm, 2017).

The use of microplastic taxa alone to identify the source of microplastics has recently been challenged (Leslie *et al.*, 2017). The two most commonly encountered taxa ascribed to a wastewater origin in our review were microbeads and fibers (Figure 8.3). Microbeads and fibers are shapes that have been found to be abundant in the environment and common in wastewater effluent (Mason *et al.*, 2016). Fibers are elongated linear objects known to come from textiles (Helm, 2017) and are also abundant in atmospheric fallout (Dris *et al.*, 2016), potentially rendering fibers an unreliable source allocation tool for wastewater. Similarly, microbeads are spherical to irregular shaped plastic objects which originate from consumer products (facial washes, cosmetics and toothpastes) – reaching the environment through wastewater effluent – and from sandblasting media (Castañeda *et al.*, 2014; Eriksen *et al.*, 2013; Free *et al.*, 2014; Gallagher *et al.*, 2016; Smith *et al.*, 2017). Size and density have aided in identifying the sources for each of these taxonomies. Microbeads are said to float if they are from a wastewater origin (cosmetic products) and sink if they are from sandblasting (Eriksen *et al.*, 2013). However, Castañeda *et al.* (2014) found non-floating microbeads that they attributed to wastewater, based on the mean diameter of the beads being similar to those found in cosmetic products. Similarly, fiber size was used by Dris *et al.* (2015) to attribute the origin of the fibers they found in stream samples to an atmospheric origin because the lengths were more similar to their atmospheric samples than to their wastewater effluent samples. By comparing fibers in wastewater effluent to environmental samples, Browne *et al.* (2011) suggested that the fingerprint from the proportions of the polymer types they found in sediments was similar to the fingerprint of fibers from laundry effluent. In this way, the strength of the evidence is amplified using multiple taxonomic characteristics beyond merely attributing all fibers or microbeads from environmental samples to wastewater. It is apparent that a thorough study of wastewater-derived microbead and fiber characteristics would be a significant contribution to the field.

8.3.2 Macroplastic indicators

Parallel to the discussion on using small microplastic taxonomies to locate sources, macroplastics (GESAMP, 2015) are distinguishable to the naked eye and can aid the identification of a microplastic source. Macroplastics can become microplastics and are often used as indicators of plastic source. Macroplastics from wastewater outfalls often reflect trash items accumulated in storm drains or items flushed into municipal sewer systems which may bypass waste treatment facilities and travel directly into waterways. When Morritt *et al.* (2014) sampled submerged litter in the river Thames (UK), they observed sanitary items and abundant litter near wastewater treatment plants, suggesting a higher proportion of plastic taxa and count can be geographically associated with wastewater outfalls. On beaches in the Bristol Channel (UK), Williams and Simmons (1997) reported macroplastics that could be assigned to wastewater outflows (sanitary items) and they attributed this to combined sewage overflow (CSO) systems whereby untreated sewage and stormwater is released to the environment during high runoff events. Similar results were reported from other beaches in the UK and elsewhere (Ross *et al.*, 1991; Storrier *et al.*, 2007; Velander & Mocogni, 1998). Since wastewater treatment technology and coverage has improved during the last decades, the occurrence of these items on beaches has decreased (Williams *et al.*, 2014), although the problem of improperly treated wastewater seems to persist even in industrialized countries (Axelsson & van Sebille, 2017), contributing significant amounts of micro- and macroplastics to rivers and the marine environment (e.g. Lahens *et al.*, 2018).

The following examples highlight the potential utility of assessing the spatial distribution of macroplastic types to identify provenance. In 2016 on an expedition to the North Atlantic Subtropical Gyre, the 5 Gyres Institute collected

38 samples with a neuston net from the sea surface between the Bahamas and New York City (5 Gyres Institute, Personal Communication 2018). The last sample was collected from the Hudson River, where a 60-minute tow in the shadow of New York City netted more plastic by weight than all the other 37 samples combined. The items were clearly associated with CSO, including plastic sticks from earbuds, tampon applicators, condoms, cigarette filters and plastic toothpicks (Figure 8.5a). Also collected were over 400 pre-production

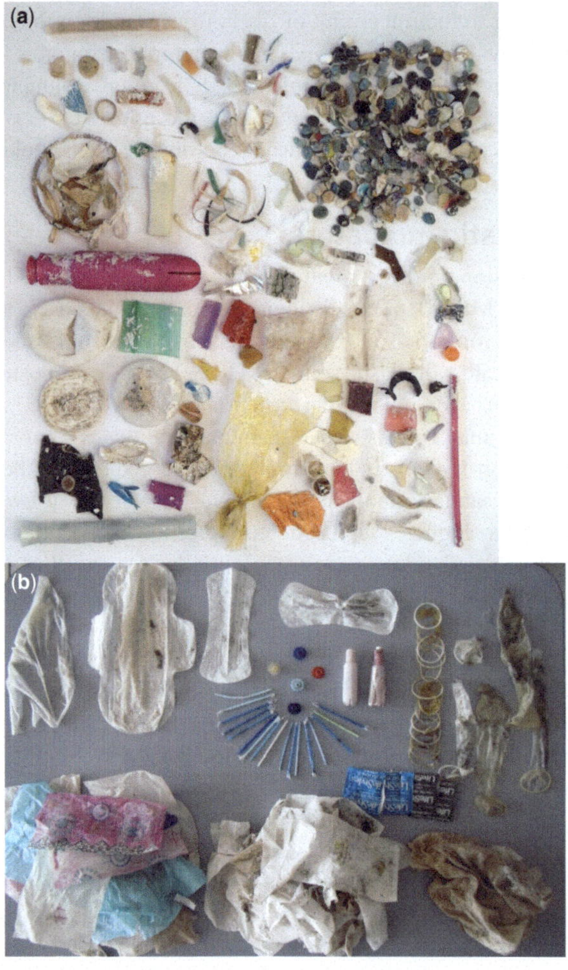

Figure 8.5 (a) results of a 60-minute surface tow in the Hudson River, showing items discharged from CSO (Photo credit: Marcus Eriksen); (b) another example of macro debris associated with wastewater discharge (Photo credit: Martin Thiel).

plastic pellets. All of these items were caked with bentonite, a clay mineral commonly used to enhance flocculation for the removal of fine particles through sedimentation. In another case, along the coast of Chile near Coquimbo, similar types of sanitary products were collected from the shore adjacent to a submarine wastewater outflow (Figure 8.5b). Though an abundance of hygiene products certainly suggests a likely wastewater source, such an approach does not produce an absolute quantitative estimation of wastewater contributions to the total population of sampled macroplastics.

8.4 WASTEWATER INDICATORS

Non-plastic indicators can help to strengthen the evidence for a microplastic source. McCormick *et al.* (2014) found microplastics in the streams they studied and used two additional forms of evidence to determine the source of the microplastics. Elevated levels of nutrients signaled an input of wastewater that corresponded to elevated levels of microplastics (McCormick *et al.*, 2014). Additionally, microbial assemblages on the microplastics were similar to those associated with wastewater (McCormick *et al.*, 2014). Talvitie *et al.* (2015) found snail shells in environmental samples that were also common in their wastewater effluent samples and concluded that the source of the microplastics was wastewater effluent. Additional indicators of wastewater that could be used in accordance with microplastic sampling are chemicals commonly used or produced as biproducts of wastewater treatment processes, such as ethylenediamintetraacetic acid, nitrilotriacetic acid, alkylphenolethoxy carboxylates, and haloacetic acids (Ding *et al.*, 1999). While multiple taxa increase certainty in source apportionment, measuring the correlation between taxa abundances and effluent may serve as further evidence.

8.5 CORRELATION

Investigating the relationship between microplastic abundance and wastewater discharge can provide a quantitative test of the hypothesis that microplastics are being introduced by wastewater effluent (Baldwin *et al.*, 2016). There are two strategies for correlating wastewater effluent to microplastic concentrations: one, based on proximity to wastewater discharge (Campbell *et al.*, 2017; Estahbanati & Fahrenfeld, 2016; Magnusson & Noren, 2014; McCormick *et al.*, 2014; Miller *et al.*, 2017; Smith *et al.*, 2017; Vermaire *et al.*, 2017) and the other on the quantity of wastewater discharged (Baldwin *et al.*, 2016; Warrack *et al.*, 2018). To date, studies applying these techniques have not addressed potential confounding factors present in their correlations.

Sampling in proximity to wastewater effluent is typically stratified by the directional fluxes of the matrix studied, as with discharge to streams (Estahbanati *et al.*, 2016) (Figure 8.6). Monitoring sites may be located above and below the

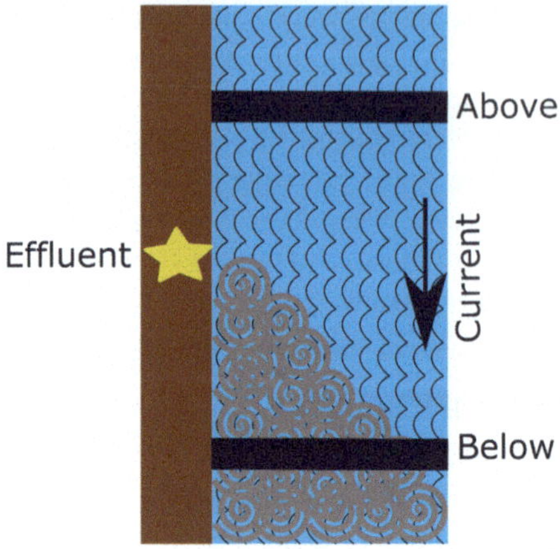

Figure 8.6 Generic sampling plan for assessing wastewater effluent impacts on microplastic concentrations in a linear flow system like a stream. Lines at "Above" and "Below" indicate generalized transect lines and represent sampling locations for a system where effluent plume structure is not known.

effluent outfall (McCormick *et al.*, 2014) or over increasing distance downstream from the outfall (Smith *et al.*, 2017). If concentrations are higher downstream or with greater proximity to the outfall, wastewater is inferred as the source. The benefit of this approach is that discharge from the effluent does not have to be known (which can be challenging data to acquire and assess) but sample sites must be stratified by effluent location.

To correlate wastewater discharge to microplastic concentration, Baldwin *et al.* (2016) sampled watershed outlets to the United States Great Lakes and, using Spearman rank procedures, did not find a significant correlation between plastic concentration and the percentage of the total streamflow from wastewater effluent. Conversely, Warrack *et al.* (2018) found that the highest season of wastewater discharge contribution corresponded with the highest concentration of microplastics found. This approach has advantages because the sites do not have to be stratified above and below the effluent pipe, but it also requires sampling a range of effluent contributions to adequately assess correlation. However, such approaches require a number of complicating assumptions, detailed below.

Complications of the proximity approach include potential issues with effluent plume mixing characteristics, confounding interactions between variables that can negatively impact the utility of correlation approaches, and the absence of flux-based considerations. While streamflow is predominantly unidirectional, the distance required for a wastewater effluent plume to fully mix across the flow field depends on the geomorphic and hydraulic conditions of the stream, the location of the effluent outfall relative to the stream (e.g. at the bank or thalweg), and differences in the physical and chemical characteristics of the effluent and river water (Roberts & Webster, 2002). Next, confounding factors from other population driven variables are likely to be present in the correlation's signal. Wastewater quantity may positively scale with population size, development intensity (Baldwin *et al.*, 2016), or tourist seasons confounding the signal from the wastewater effluent with other potential sources of microplastics in the region. These potential confounding factors have yet to be explored explicitly. The correlation between wastewater proportional quantities such as percentage flow contribution (Baldwin *et al.*, 2016) may avoid these confounding factors when the connectivity between the wastewater outfall and the sampling location can be assumed equal among studied sites. Additionally, investigating correlations between microplastic characteristic taxa like microbead, fiber and fragment concentrations could provide more information than derived from correlative analyses of bulk concentrations alone (Baldwin *et al.*, 2016). However, potential exchange of microplastics with channel banks and beds may complicate even simple cases of stratified outfall sampling in a channelized system (Klein *et al.*, 2015). Furthermore, investigation of concentration without corresponding water discharge data omits the possibility of estimating absolute microplastic mass flux from effluent to receiving bodies, which may be present despite relative dilution from the effluent.

8.6 MASS BALANCE

The most rigorous approach to quantifying the impact of wastewater effluent on the abundance and character of microplastics in an aquatic system is a complete microplastic mass balance. To date, no studies have used this method. The components of a generic mass balance are: (i) identifying the boundaries of the aquatic system of interest; (ii) determining which boundaries are relevant to sample; and (iii) measuring or estimating the flux of microplastics across each boundary (Edwards & Glysson, 1999). Here we discuss the application of the mass balance approach to a river or stream setting (which are the most common systems studied in this review), but the approach can be adapted to other systems.

The microplastic boundary conditions of a given stream include at least the flux of waterborne microplastics from upstream, the flux of wastewater effluent microplastics and the efflux of microplastics out of the stream reach. However, additional boundaries that may serve as sources or sinks include the channel bed and banks, other surface water compartments and the atmosphere. Microplastics in atmospheric fallout are common and a likely source of contamination in samples (Dris *et al.*, 2016). Erosion or aggradation of stream bed and bank material can release or sequester microplastics to or from the flow field (Besseling *et al.*, 2017) but, even in cases of stable bed elevation, exchange of microplastic material likely occurs (Walling *et al.*, 1998). The first step toward better understanding the communication of microplastics between stream flow and channel bed and bank materials must include further study of changes of in-channel microplastic storage over time, including spatio-temporal details of aggradational/degradational processes.

Choosing which boundaries are important requires prior knowledge of the various sources. In most cases potential sources and sinks could be reasonably assumed to be insignificant, thereby simplifying the mass balance scheme. For example, concrete drainage canals retaining little to no sediment could be assumed to have no bed and bank exchange of microplastics, an assumption that may also be applicable to "natural" channels found to be in dynamic equilibrium over the course of a study. Aeolian fluxes over the areas and time scales of sampling may also be found to be inconsequential. If the upstream channel length is much longer than the study reach, one would expect that a given parcel of water would be exposed to much more atmospheric fallout of microplastics during the travel time to, rather than through, the study reach. In most stream mass balance scenarios, one would expect that the flux of microplastics from upstream and from the wastewater effluent would be the most important components for assessing the importance of the wastewater contribution.

Spatio-temporal dynamics of particle transport and study constraints can impact decisions about how to measure microplastics. The geomorphology and hydrology of the stream channel can greatly affect the concentrations of suspended particles over short distances or times (Walling, 1983). Eddy currents in a stream can concentrate particles, and turbulent fluctuations can carry bursts of sediment and potentially denser microplastics to the surface (where most microplastic sampling has taken place) (Gray & Gartner, 2010). A comprehensive water sampling scheme should seek to dampen these short-term/range variations through samples that integrate over time periods and distances long enough to remove potential bias and outliers. The large sample requirement for microplastic analysis (often on the order of cubic meters) necessitates longer sampling times, and such considerations provide additional support for cross-channel sampling transects (Figure 8.6). Quantifying microplastic storage adjacent to the system and the flux across system boundaries imposes additional logistical constraints. Some fluxes, like aeolian microplastic deposition rates, may be relatively easily monitored with deposition pans (Dris *et al.*, 2016). However, measuring microplastic flux to/from the channel bed and banks is challenging and requires prior knowledge about the depositional morphology of the location (Hurley *et al.*, 2018). Obtaining effluent water flux and microplastic abundance and character from the wastewater source itself would be ideal and could make a mass balance unnecessary when answering the question "How much microplastic is coming from the effluent?"

8.7 STANDARDIZATION

How should source allocation of microplastics to wastewater be standardized? The reviewed studies nearly all monitored aquatic systems by sampling only the top of the water column, and in some cases channel bed sediments, leaving most of the stream water column and channel banks unmonitored. Sample sizes ranged from 1 L grab samples (Miller *et al.*, 2017) to long trawls through many cubic meters of water (Eriksen *et al.*, 2013), with smaller samples generally resulting in much higher concentrations (Barrows *et al.*, 2017). Minimum particle size thresholds ranged from 5–500 µm and there were 20 unique microplastic size ranges introduced in the reviewed literature. To merge these data requires a number of assumptions about the total size distribution of microplastics sampled in each study. Only two of the studies quantified microplastic mass (Free *et al.*, 2014; Lechner *et al.*, 2014); the rest measured count alone. However, the range of error in converting between count and mass could be as high as five orders of magnitude (Schmidt *et al.*, 2017) (Figure 8.7). If researchers measured the size of each particle directly, scientists could be able to more readily and accurately compare results (Mintenig *et al.*, 2018). Lechner *et al.* (2014) compared mass to count and found that the proportional abundance of shape taxonomies changed, consequently redistributing the rank of the taxonomies – which begs the question: should researchers be using count (which is not a conserved unit) to measure flux? Furthermore, access to data is a requirement for repeating results and comparing literature; however, our investigation indicates that only six of the reviewed papers had published data through an open access portal by the date of this review. To standardize future research, sampling and analysis protocols should be developed for the full water column and stream bank; additionally, efforts toward quantifying mass, count, and particle size characteristics should be emphasized, and adopting a community open access policy for data archiving and dissemination should be prioritized.

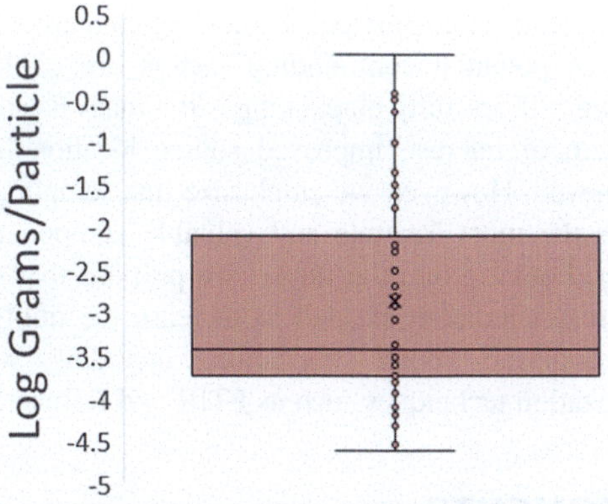

Figure 8.7 The log normalized grams per particle extracted from stream microplastics studies in Schmidt *et al.* (2017), showing that there is a range of five orders of magnitude involved in converting between particle count and mass.

After reviewing these recent papers on microplastic pollution from wastewater effluent, we must address a fundamental question: are these studies reliable? The majority of them (14) employed no chemical confirmation to guarantee plastic as the material and relied on visual observation alone. The smaller the particle, the less accurate the plastic identification by visual techniques, with particles below 1.5 mm having a significant increase in error of identification (Löder *et al.*, 2017; Kroon *et al.*, 2018). Raman and FTIR spectroscopy are complimentary techniques for determining microplastic polymer types, as they are able to characterize extremely small microplastics (with a minimum size of 1 μm and 20 μm, respectively) (Käppler *et al.*, 2016). The cost and time of analysis is substantial (in our experience ~$500–1000 and ~10–50 h per sample), and it is evident that automated techniques will be required in the future (Primpke *et al.*, 2018). However, the field of microplastic pollution is rapidly moving in the direction of spectral verification and we expect that wastewater studies employing analytical chemistry techniques will become the norm.

8.8 CONCLUSIONS

This chapter has organized existing approaches into a framework that can be used in a multiple lines of evidence approach to assess the source of microplastics in the environment. Some forms of evidence should be given more weight than others. While anecdotal evidence provides a reason for further investigation, it should be followed by investigation founded on more quantitative techniques. Taxonomic evidence can define a fingerprint that may be matched to a source but, as yet, there is no accepted or standardized taxonomic system or strategy for microplastics. For now, multiple forms of taxonomic evidence are essential to any source identification. Other forms of nonplastic evidence related to water sources should be combined with information on the various environmental forces acting on the sample area. Correlation can provide a more quantitative source attribution technique as long as potential confounding factors are explored in depth. By explicitly considering – if not fully elucidating – the mass balance of microplastics in the aquatic system of interest, improved source location from environmental samples can be achieved. However, we emphasize that sampling the effluent itself, if possible, remains the most accurate and valuable component of a wastewater source investigation. Looking into the future, we propose that standardization and validation efforts are immediately needed to increase the utility and reliability of environmental microplastic source allocation, including wider adoption of molecular characterization techniques such as FTIR and Raman Spectroscopy.

ACKNOWLEDGMENTS

This study was funded in part by the USDA National Institute of Food and Agriculture, Hatch program (A. Gray, project number CA-R-ENS-5120-H) and a

National Science Foundation Graduate Research Fellowship (W. Cowger). We are thankful for the valuable input from a peer reviewer. We thank Kevin Yan, Noelani Leal, and Kristen Briseno who provided valuable input toward the development of the ideas and questions expressed in this chapter.

REFERENCES

Axelsson C. and van Sebille E. (2017). Prevention through policy: Urban macroplastic leakages to the marine environment during extreme rainfall events. *Marine Pollution Bulletin*, **124**(1), 211–22.

Baldwin A. K., Corsi S. R. and Mason S. A. (2016). Plastic debris in 29 great lakes tributaries: relations to watershed attributes and hydrology. *Environmental Science & Technology*, **50**(19), 10377–10385.

Barrows A. P. W., Neumann C. A., Berger M. L. and Shaw S. D. (2017). Grab vs. neuston tow net: a microplastic sampling performance comparison and possible advances in the field. *Analytical Methods*, **9**(9), 1446–1453.

Besseling E., Quik J. T. K., Sun M. and Koelmans A. A. (2017). Fate of nano- and microplastic in freshwater systems: A modeling study. *Environmental Pollution*, **220** (Pt A), 540–548.

Blair R. M., Waldron S., Phoenix V. and Gauchotte-Lindsay C. (2017). Micro- and nanoplastic pollution of freshwater and wastewater treatment systems. *Springer Science Reviews*, **5**(1-2), 19–30. http://dx.doi.org/10.1007/s40362-017-0044-7

Browne M. A., Crump P., Niven S. J., Teuten E., Tonkin A., Galloway T. and Thompson R. (2011). Accumulation of microplastic on shorelines worldwide: sources and sinks. *Environmental Science & Technology*, **45**(21), 9175–9179.

Campbell S. H., Williamson P. R. and Hall B. D. (2017). Microplastics in the gastrointestinal tracts of fish and the water from an urban prairie creek. *FACETS* **2**(1), 395–409.

Carr S. A., Liu J. and Tesoro A. G. (2016). Transport and fate of microplastic particles in wastewater treatment plants. *Water Research*, **91**, 174–182.

Castañeda R., Suncica Avlijas M., Simard A. and Ricciardia A. (2014). Microplastic pollution discovered in St. Lawrence River sediments. *NRC Press*, **88**(1–2), 5–6.

CA State Legislature (2015). Waste Management: Plastic Microbeads. *Assembly Bill No. 888 Chapter 594*. See: https://leginfo.legislature.ca.gov/faces/billNavClient.xhtml?bill_id=201520160AB888 (accessed 17 February 2018).

Claessens M., De Meester S., Van Landuyt L., De Clerck K. and Janssen C. R. (2011). Occurrence and distribution of microplastics in marine sediments along the Belgian coast. *Marine Pollution Bulletin*, **62**(10), 2199–2204.

Ding W. H., Wu J., Semadeni M. and Reinhard M. (1999). Occurrence and behavior of wastewater indicators in the Santa Ana River and the underlying aquifers. *Chemosphere*, **39**(11), 1781–1794.

Dris R., Gasperi J., Saad M., Mirande C. and Tassin B. (2016). Synthetic fibers in atmospheric fallout: A source of microplastics in the environment? *Marine Pollution Bulletin*, **104** (1–2), 290–293.

Dris R., Gasperi J., Rocher V., Saad M., Renault N. and Tassin B. (2015). Microplastic contamination in an urban area: a case study in Greater Paris. *Environmental Chemistry*, **12**(5), 592–599.

Dubaish F. and Liebezeit G. (2013). Suspended microplastics and black carbon particles in the jade system, Southern North Sea. *Water, Air, & Soil Pollution: Focus*, **224**(2), 1352.

Edwards T. K. and Glysson D. (1999). *Field Methods for Measurement of Fluvial Sediment. in Techniques of Water-Resources Investigations of the U.S. Geological Survey. Book 3, Applications of Hydraulics. Chapter C2.* USGS Reston, VA, 89 pp.

Eriksen M., Mason S., Wilson S., Box C., Zellers A., Edwards W., Farley H. and Amato S. (2013). Microplastic pollution in the surface waters of the Laurentian Great Lakes. *Marine Pollution Bulletin*, **77**(1–2), 177–182.

Estahbanati S. and Fahrenfeld N. L. (2016). Influence of wastewater treatment plant discharges on microplastic concentrations in surface water. *Chemosphere*, **162**, 277–284.

Fendall L. S. and Sewell M. A. (2009). Contributing to marine pollution by washing your face: microplastics in facial cleansers. *Marine Pollution Bulletin*, **58**(8), 1225–1228.

Free C. M., Jensen O. P., Mason S. A., Eriksen M., Williamson N. J. and Boldgiv B. (2014). High-levels of microplastic pollution in a large, remote, mountain lake. *Marine Pollution Bulletin*, **85**(1), 156–163.

Gago J., Carretero O., Filgueiras A. V. and Viñas L. (2018). Synthetic microfibers in the marine environment: A review on their occurrence in seawater and sediments. *Marine Pollution Bulletin*, **127**, 365–376.

Gallagher A., Rees A., Rowe R., Stevens J. and Wright P. (2016). Microplastics in the Solent estuarine complex, UK: An initial assessment. *Marine Pollution Bulletin*, **102**(2), 243–249.

GESAMP (2015). Sources, fate and effects of microplastics in the marine environment: a global assessment. In: (IMOFAOUNESCO–IOCUNIDOWMOIAEAUNUNEP UNDP Joint Group of Experts on the Scientific Aspects of Marine Environmental Protection (GESAMP)), P. J. Kershaw (ed.), *Rep. Stud. GESAMP No. 90*, 96p.

Gray J. R. and Gartner J. W. (2010). Overview of Selected Surrogate Technologies for High-temporal Resolution Suspended-Sediment Monitoring. Proceedings of the 2nd Joint Federal Interagency Conference, Las Vegas, NV.

Gregory M. R. (1996). Plastic "scrubbers" in hand cleansers: a further (and minor) source for marine pollution identified. *Mar Pollut Bull*, **32**(12), 867–871.

Habib D., Locke D. C. and Cannone L. J. (1998). Synthetic fibers as indicators of municipal sewage sludge, sludge products, and sewage treatment plant effluents. *Water, Air, & Soil Pollution*, **103**(1), 1–8.

Hanvey J. S., Lewis P. J., Lavers J. L., Crosbie N. D., Pozo K. and Clarke B. O. (2017). A review of analytical techniques for quantifying microplastics in sediments. *Analytical Methods*, **9**(9), 1369–1383.

Helm P. A. (2017). Improving microplastics source apportionment: a role for microplastic morphology and taxonomy? *Analytical Methods*, **9**(9), 1328–1331.

Hidalgo-Ruz V., Gutow L., Thompson R. C. and Thiel M. (2012). Microplastics in the marine environment: a review of the methods used for identification and quantification. *Environmental Science and Technology*, **46**(6), 3060–75.

Hurley R., Woodward J. and Rothwell J. J. (2018). Microplastic contamination of river beds significantly reduced by catchment-wide flooding. *Nature Geoscience*, **11**(4), 251–257.

Käppler A., Fischer D., Oberbeckmann S., Schernewski G., Labrenz M., Eichhorn K. J. and Voit B. (2016). Analysisof environmental microplastics by vibrational microspectroscopy: FTIR, Raman or both? *Analytical and Bioanalytical Chemistry*, **408**(29), 8377–8391.

Klein S., Worch E. and Knepper T. P. (2015). Occurrence and spatial distribution of microplastics in river shore sediments of the Rhine-main area in Germany. *Environmental Science and Technology*, **49**(10), 6070–6076.

Kroon F., Motti C., Talbot S., Sobral P. and Puotinen M. (2018). A workflow for improving estimates of microplastic contamination in marine waters: A case study from North-Western Australia. *Environmental Pollution*, **238**, 26–38.

Lahens L., Strady E., Kieu-Le T. C., Dris R., Boukerma K., Rinnert E., Gasperi J. and Tassin B. (2018). Macroplastic and microplastic contamination assessment of a tropical river (Saigon River, Vietnam) transversed by a developing megacity. *Environmental Pollution*, **236**, 661–671.

Lechner A., Keckeis H., Lumesberger-Loisl F., Zens B., Krusch R., Tritthart M., Glas M. and Schludermann E. (2014). The Danube so colourful: a potpourri of plastic litter outnumbers fish larvae in Europe's second largest river. *Environmental Pollution*, **188**, 177–81.

Leslie H. A., Brandsma S. H., van Velzen M. J. M. and Vethaak A. D. (2017). Microplastics en route: Field measurements in the Dutch river delta and Amsterdam canals, wastewater treatment plants, North Sea sediments and biota. *Environment International*, **101**, 133–42.

Li J., Liu H. and Chen J. P. (2017). Microplastics in freshwater systems: a review on occurrence, environmental effects, and methods for microplastics detection. *Water Research*, **137**, 362–374. https://doi.org/10.1016/j.watres.2017.12.056. See: http://www.sciencedirect.com/science/article/pii/S0043135417310515 (accessed 12 May 2019).

Löder M. G. J., Imhof H. K., Ladehoff M., Löschel L. A., Lorenz C., Mintenig S., Piehl S., Primpke S., Schrank I., Laforsch C. and Gerdts G. (2017). Enzymatic purification of microplastics in environmental samples. *Environmental Science and Technology*, **51** (24), 14283–92.

Magnusson K. and Norén F. (2014). Screening of microplastic particles in and down-stream a wastewater treatment plant. Swedish Environmental Research Institute, Stockholm, p. 22.

Mai L., Bao L.-J., Shi L., Wong C. S. and Zeng E. Y. (2018). A review of methods for measuring microplastics in aquatic environments. *Environmental Science and Pollution Research* **25**(12), 11319–11332.

Mason S. A., Garneau D., Sutton R., Chu Y., Ehmann K., Barnes J., Fink P., Papazissimos D. and Rogers D. L. (2016). Microplastic pollution is widely detected in US municipal wastewater treatment plant effluent. *Environ Pollut*, **218**, 1045–54.

McCormick A., Hoellein T. J., Mason S. A., Schluep J. and Kelly J. J. (2014). Microplastic is an abundant and distinct microbial habitat in an urban river. *Environmental Science and Technology*, **48**(20), 11863–71.

Miller R. Z., Watts A. JR., Winslow B. O., Galloway T. S. and Barrows A. P. W. (2017). Mountains to the sea: River study of plastic and non-plastic microfiber pollution in the northeast USA. *Marine Pollution Bulletin*, **124**(1), 245–251. https://doi.org/10.1016/j.marpolbul.2017.07.028.

Mintenig S. M., Bäuerlein P. S., Koelmans A. A., Dekker S. C. and van Wezel A. P. (2018). Closing the gap between small and smaller: towards a framework to analyse nano- and microplastics in aqueous environmental samples. *Environmental Science: Nano*, **5**(7), 1640–1649.

Morritt D., Stefanoudis P. V., Pearce D., Crimmen O. A. and Clark P. F. (2014). Plastic in the Thames: a river runs through it. *Marine Pollution Bulletin*, **78**(1), 196–200.

Pizzuto J., Keeler J., Skalak K. and Karwan D. (2017). Storage filters upland suspended sediment signals delivered from watersheds. *Geology*, **45**(2), 151–154.

Primpke S., Wirth M., Lorenz C. and Gerdts G. (2018). Reference database design for the automated analysis of microplastic samples based on Fourier transform infrared (FTIR) spectroscopy. *Analytical and Bioanalytical Chemistry*, **410**(21), 5131–5141. http://dx.doi.org/10.1007/s00216-018-1156-x.

Roberts P. J. W. and Webster D. R. (2002). Turbulent diffusion. In: Environmental Fluid Mechanics: Theories and Application, H. H. Shen (ed.), American Society of Civil Engineering, New York.

Ross J. B., Parker R. and Strickland M. (1991). A survey of shoreline litter in Halifax Harbour 1989. *Marine Pollution Bulletin*, **22**(5), 245–248.

Schmidt C., Krauth T. and Wagner S. (2017). Export of Plastic Debris by Rivers into the Sea. *Environmental Science & Technology*, **51**(21), 12246–12253. http://dx.doi.org/10.1021/acs.est.7b02368.

Shim W. J., Hong S. H. and Eo S. E. (2017). Identification methods in microplastic analysis: a review. *Anal Methods*, **9**(9), 1384–91.

Silva A. B., Bastos A. S., Justino C. I. L., da Costa J. P., Duarte A. C. and Rocha-Santos T. A. P. (2018). Microplastics in the environment: challenges in analytical chemistry – a review. *Anal Chim Acta*, **1017**, 1–19. See: http://www.sciencedirect.com/science/article/pii/S0003267018302587 (accessed 12 May 2019).

Smith J. A., Hodge J. L., Kurtz B. H. and Garver J. I. (2017). The Distribution of Microplastic Pollution in the Mohawk River. Mohawk Watershed Symposium.

Storrier K. L., McGlashan D. J., Bonellie S. and Velander K. (2007). Beach litter deposition at a selection of beaches in the Firth of Forth, Scotland. *Journal of Coastal Research*, 813–822.

Talvitie J., Heinonen M., Pääkkönen J. P., Vahtera E., Mikola A., Setälä O. and Vahala R. (2015). Do wastewater treatment plants act as a potential point source of microplastics? Preliminary study in the coastal Gulf of Finland, Baltic Sea. *Water Science and Technology*, **72**(9), 1495–504.

US Congress (2015). HR 1321 Microbead-Free Waters Act of 2015. See: https://www.congress.gov/bill/114th-congress/house-bill/1321/text/rds (accessed 17 February 2018).

Velander K. A. and Mocogni M. (1998). Maritime litter and sewage contamination at Cramond Beach Edinburgh – a comparative study. *Marine Pollution Bulletin*, **36**(5), 385–389.

Vermaire J. C., Pomeroy C., Herczegh S. M., Haggart O. and Murphy M. (2017). Microplastic abundance and distribution in the open water and sediment of the Ottawa River, Canada, and its tributaries. *FACETS* **2**(1), 301–14.

Walling D. E. (1983). The sediment delivery problem. *J. Hydrol*, **65**(1), 209–237.

Walling D. E., Owens P. N. and Leeks G. J. L. (1998). The role of channel and floodplain storage in the suspended sediment budget of the River Ouse, Yorkshire, UK. *Geomorphology.*, **22**(3–4), 225–242.

Warrack S., Challis J. K., Hanson M. L. and Rennie M. D. (2018). Microplastics Flowing into Lake Winnipeg: Densities, Sources, Flux, and Fish Exposures. Proceedings of Manitoba's Undergraduate Science and Engineering Research.

Williams A. T. and Simmons S. L. (1997). Estuarine Litter at the River/Beach Interface in the Bristol Channel, United Kingdom. *Journal of Coastal Research*, **13**(4), 1159–1165.

Williams A. T., Randerson P. and Alharbi O. A. (2014). From a millennium base line to 2012: Beach litter changes in Wales. *Marine Pollution Bulletin*, **84**(1–2), 17–26.

Woodall L. C., Sanchez-Vidal A., Canals M., Paterson G. L. J., Coppock R., Sleight V., Calafat A. and Rogers A. D. (2014). The deep sea is a major sink for microplastic debris. *Royal Society Open Science*, **1**(4), 140317.

Zhang K., Gong W., Lv J., Xiong X. and Wu C. (2015). Accumulation of floating microplastics behind the Three Gorges Dam. *Environ Pollut*, **204**, 117–123.

Zhao S., Zhu L., Wang T. and Li D. (2014). Suspended microplastics in the surface water of the Yangtze Estuary System, China: first observations on occurrence, distribution. *Marine Pollution Bulletin*, **86**(1–2), 562–568.

Ziajahromi S., Neale P. A. and Leusch F. D. L. (2016). Wastewater treatment plant effluent as a source of microplastics: review of the fate, chemical interactions and potential risks to aquatic organisms. *Water Science and Technology*, **74**(10), 2253–2269.

Chapter 9

Pollution of beaches and watercourses by plastic biomedia

P. Bencivengo and C. Barreau

Environment Division, Surfrider Foundation Europe, Biarritz, France

Keywords: Beach cleanup, Coast, Filters, Incidents, MBBRs, Marine Litter, Participatory science, Plant, Sewage

9.1 INTRODUCTION TO BIOMEDIA POLLUTION

Large numbers of small plastic cylinders have been found washed up along European coasts since 2007. These objects have been identified as bacterial biofilm carriers used in the wastewater treatment process and can be referred to as biomedia or filter media. Pollution in the form of these plastic cylinders now seems to affect every coastline in the world.

This chapter is a synthesis of a larger study (Bencivengo *et al.*, 2018) aiming to share the data that the Surfrider Foundation has gathered over the course of its seven-year investigation into biomedia pollution, to better understand how a process created to clear water ends up polluting the environment.

This study involved making information requests and conducting interviews with wastewater industry experts in order to gain an objective understanding of how biomedia use could lead to losses, and to work together to come up with workable and environmentally-friendly solutions.

Editors: Hrissi K. Karapanagioti and Ioannis K. Kalavrouziotis
doi: 10.2166/9781789061697_0133

9.2 BACKGROUND TO WASTEWATER PURIFICATION AND BIOLOGICAL TREATMENT

9.2.1 Overview of operations in a sewage treatment system

Water used by both households and industrial sites must pass through a wastewater treatment plant (WWTP) in order to protect public health, the environment and water resources. WWTPs are generally composed of a succession of physical and chemical processes: a primary treatment to remove solid material and a secondary treatment to digest dissolved and suspended organic material. A tertiary treatment can also be added, providing a disinfection before discharging water to the environment.

9.2.2 Focus on biological treatment

During the biological treatment stage, which is part of the secondary treatment, organic matter is broken down by heterotrophic bacteria. Secondary biological treatments can be divided into extensive (natural) and intensive (mechanical) processes.

9.2.2.1 Extensive biological processes

These are processes that harness the environment's own natural purification capacity. Water can be treated by a reed bed, through lagooning, by the creation of a wetland area or through percolation, none of which involve any mechanical intervention.

9.2.2.2 Intensive biological processes

These processes use bacterial cultures combined with mechanical treatment and artificial oxygenation to treat the wastewater more quickly and in limited spaces. There are two major categories of intensive biological processes:

9.2.2.3 Free culture installations: activated sludge

The bacterial culture is maintained in an aerated basin, where it is continuously mixed, facilitating the biodegradation process by keeping the bacteria in contact with the polluting particles. In this process, the purifying microorganisms clump together in flocs. This reduces the exchange surface, and consequently the system's effectiveness and performance.

9.2.2.4 Fixed-film installations

The bacteria used to break down the organic matter are grown on a variety of supports in the form of biofilms. The supports provided for the growth of this biomass mean that a larger number of cells can develop, thereby increasing the purification capacity of the installation. The activity of a bacterial culture depends primarily on the exchange surface between the biofilm and the oxygenated effluent (Canler & Perret, 2012): the greater the surface area, the greater the cleaning capacity. This area is generally indicated in m^2 of colonised surface/m^3 of the support. There are several solutions for optimising the treatments, such as trickle filters, rotating biological contactors, biological filters, fluidised bed reactors and mixed solutions.

9.2.3 Fluidised bed bioreactors

Biological treatment using fluidised bed bioreactors has heralded a technological and economic revolution in the world of wastewater treatment. This process revolves around the use of biomedia.

9.2.3.1 Principles

The aim of a fluidised bed bioreactor system, also known as a Moving Bed Biofilm Reactor (MBBR), is to provide bacteria with an environment that will allow them to develop optimally in a compact space, in order to break down the pollutants in the water. This optimisation depends on two major factors: the supports upon which the bacteria can develop and access to nutrients (Canler *et al.*, 2012).

The support is provided by the biomedia, which are made of plastic (either polyethylene or high-density polyethylene). These are added to the bioreactors at a rate of 30–65% of the volume of the basin (Canler *et al.*, 2012). This means that there are hundreds of thousands or even millions of plastic pieces in each reactor. Their honeycombed, colonisable structure and their density, which is similar to that of water (1 g/cm^3), makes it easy to keep them moving within the tank either by mechanical aeration or mixing. This movement should be uniform to ensure an optimal level of contact between the microorganisms and the effluent to be treated. This process depends upon the type of support chosen and the rate at which the treatment basins are refilled.

Biomedia can be used in different phases of the biological treatment process: pre-treatment, secondary treatment and even in combination with activated sludge (Canler *et al.*, 2012). This flexibility means the system can be a very attractive option for new WWTPs. Fluidised bed bioreactors can also be introduced during upgrades at older WWTPs. This makes it possible to increase a plant's treatment capacity without the need to build any new basins – an approach that is often heavily driven by financial or space constraints.

The parameters used to calculate the volume of biomedia needed for water treatment are incoming flow, discharge flow and effluent temperature. The optimal functioning of the wastewater treatment infrastructure therefore depends heavily on this calculation, which impacts on the whole plant's performance and ability to achieve its objectives.

9.2.3.2 Advantages

The biomedia process offers advantages such as: adaptability (Canler *et al.*, 2012; Laurent, 2006); high concentration of available biomass (Kargi & Karapinar, 1997; Nicolella *et al.*, 2000); lengthy biomass survival time (Nicolella *et al.*, 2000); improved mass transfer (Jianping *et al.*, 2003; Nicolella *et al.*, 2000; Venu Vinod & Venkat Reddy, 2005); reduced water retention time (Gonzalez *et al.*, 2001; Jianping *et al.*, 2003; Kargi & Karapinar, 1997); ease of cleaning (Kargi & Karapinar, 1997); a compact procedure (Canler *et al.*, 2012); and a higher ability to biodegrade micropollutants compared to activated sludge systems due to the growth of specified organisms on them (Mazioti *et al.*, 2015).

9.2.3.3 Limitations and disadvantages

While this process has some clear advantages, it also has inherent risks and constraints including: poor bacterial activity at low temperatures ($<5°C$); it is an energy-hungry and costly process; the slow colonisation of biomedia by the bacterial biofilms (Nicolella *et al.*, 2000); and loss of the biomedia.

9.3 USERS

MBBR systems are used today for treating wastewater in public and industrial WWTPs, as well as in individual private systems and in the farming sector.

9.3.1 Municipal sewage treatment

If a dwelling is connected to the local sewage network, it becomes part of the municipal mains wastewater treatment system, which is the most common system in urban areas. Nearly all towns with a population over 10,000 today have their own WWTP. MBBR processes can be used by communities and towns ranging in size from a few thousand to many tens of thousands of inhabitants.

9.3.2 Private off-mains sewage treatment

Unlike mains wastewater systems, off-mains wastewater treatment (also called domestic or individual systems) are facilities that are not connected to the public network. Depending on the volume of effluent to be treated, these can range from industrial WWTPs able to treat many thousands of Population Equivalents (PE) or micro-stations designed to treat much smaller volumes. In general, these systems are used to meet the challenges of isolated locations, specific arrangements (e.g. fish ponds), to treat wastewater from small industrial businesses before they are discharged to the environment, or for pre-treatment of industrial effluent prior to it being discharged into the municipal sewage network.

Other domestic facilities operated by private individuals which are unregulated such as swimming pools, natural lakes and ornamental ponds also require regular water treatment. Inspired by professional fish farms, many amateurs use biomedia to filter the water in their ponds. Unfortunately, the suppliers of these items often deliver them without any explanation of how to use them, leaving the purchasers to work out how to install and use them on a trial and error basis.

9.3.3 Non-public industrial wastewater treatment

Companies producing industrial effluent such as for paper and cardboard production, chemical wood processing, agrifoods or fish farming are subject to special measures. All industries, no matter what they produce, are obliged to treat their effluent. Industrial effluent may then be discharged back into the environment either after treatment by the business itself (independent treatment), or after being discharged into the municipal sewage network.

Treating industrial wastewater is a complex matter. Each facility is a different case, which needs its own specially-adapted equipment and processes in order to fulfil its requirements. Strict environmental constraints, protections and the large volumes of water involved in industrial processes mean companies must adopt methods to limit their water consumption and encourage water recycling.

9.4 THE SPREAD OF BIOMEDIA IN THE NATURAL ENVIRONMENT

Biomedia spread through the environment if they escape from WWTPs, firstly through freshwater systems and then to the sea. Some of them will end up being washed up on the coast (Figure 9.1), sometimes thousands of kilometres from their source (Bencivengo *et al.*, 2018). To understand how they spread, it is essential to understand the environmental, weather and water-related factors that interact with these items of floating debris.

Figure 9.1 Biomedia washed up on a beach, Aquitaine, France.

9.4.1 Land-based sources and transportation in waterways

Biomedia escaping water treatment plants can, like any unnatural element entering the environment, end up in the sea. They can be transported in water courses over hundreds of kilometres from their point of discharge, just as a drop of water will also follow the same route through the water cycle. This means biomedia can be dispersed over vast areas.

9.4.1.1 The upstream–downstream connection

It is estimated that 80% of all the waste found on our coasts has a land-based source (Araújo & Costa, 2007; Jambeck *et al.*, 2015). The main vectors for the spread of pollution from inland areas to the oceans are rivers. WWTPs generally discharge into water courses and this is therefore the principal means by which biomedia are lost into the environment. Rainfall impacts on water levels and river flows. The ebb and flow between low- and high-water levels affects how a water course is able to remobilise waste deposited on its banks. Once they are picked up by the rivers, these waste items follow their route downstream. Estuaries mark the interface between freshwater and salt water, and it is here, at river mouths, that waste finds itself flowing out into the marine environment.

9.4.1.2 Currents

The world's oceans are in a state of perpetual motion, thanks to the forces acting on water masses (winds, tides, Coriolis force) and their physical-chemical properties. From river mouths, waste can be transported many thousands of kilometres by surface currents.

9.5 MONITORING BIOMEDIA POLLUTION

In 2007, a volunteer with Surfrider Foundation Europe started to notice biomedia on the beaches of the French Basque coast. Over the years, these media started to turn up along all French and European coasts. Surfrider Foundation Europe has gained significant expertise and become the leading organisation working on this issue, thanks to its extensive network and the data collected by a network of external observers.

More and more reports (over 500) were thus collected, with the participation of other European NGOs and ocean clean up organisations. The wide spread of identification sheets enabled qualitative and quantitative data to be gathered about the biomedia found along the coasts, and therefore to establish trends relating to its concentration.

Today, most of the leading firms working in the wastewater industry have adopted the moving bed process and developed their own models of plastic carrier. Each type of biomedia has a different shape and surface area and is designed for a particular purpose, making the biomedia a specific factor at each plant. This also means that biomedia can be used to track uses and processes and can be traced back to their source if they are found in the environment.

9.6 BIOMEDIA POLLUTION INCIDENTS

Numerous biomedia pollution incidents have been reported since 2007 along large stretches of European rivers and coasts. Follow-up investigations have been conducted at some of the sites suffering the most serious impacts, with a view to establishing the source of the discharge. Two of these are presented below.

9.6.1 Saint-Prex (Switzerland)

9.6.1.1 General information

- Region: City of Saint-Prex – Canton of Vaud (Switzerland)
- Affected waters: Lake Geneva
- Plant: Joint municipal plant of Saint-Prex, Etoy and Buchillon. Started operating in 1977 and equipped with biomedia since April 2012
- Nominal capacity: 16,000 PE
- Type of biomedia found: BWT 15

9.6.1.2 Account of the incident

The amount of water entering the plant increased dramatically following a violent storm on 17/18 September 2012. Manual attempts to use an overflow channel to reduce the influent flow caused a wave in the settlement tank. This resulted in the biomedia being pushed towards the water exit mesh, which became blocked, leading to the basin overflowing (Figure 9.2).

Figure 9.2 Loss of Biomedia at the outflow of a WWTP, Switzerland.

A second failure occurred around the oxygen sensors in the aeration basin. The aerated storm water disrupted the sensors, which then sent a signal for the air supply in the tank to be reduced, which exacerbated the blockage. A subsequent investigation by a volunteer also showed that neither the plant's management nor the local authority followed the cantonal alert procedure.

9.6.1.3 Actions undertaken

No steps were taken to contain the pollution during the storm event. However, technical modifications have subsequently been made to the basins:

- the diversion channel has been modified, which today responds automatically to changes in the influent flow at the WWTP;
- water level sensors have been put in place, making it possible to detect different water levels and reduce the incoming flow. This system allows the diffusers to be overridden to increase air input and prevent the meshes from becoming blocked;
- perforated stainless steel tubes welded horizontally to the outflow mesh have been installed, enabling water to continue to pass through even in the event of blockages.

Both the company that installed the system and the supplier of the biomedia have reacted following this pollution incident, now having precautions in their protocols in order to prevent any further incidents at the WWTP. In December 2013, over a year after the incident, the town council of Saint-Prex reported on the technical improvements put in place following the spill to prevent any further problems of this kind.

9.6.1.4 Results

Thousands of biomedia were collected from all around the banks of Lake Geneva. Although they are less common today, the biomedia still continue to wash up all around the lake, demonstrating the significant environmental impact of this kind of pollution. The same biomedia, so characteristic of the pollution event in Lake Geneva, have also been found during *Ocean Initiatives* clean-up events on the coasts of the Mediterranean, showing once again just how far this pollution can spread, and the key role that the river system plays in dispersing biomedia.

After being judged by local authorities to be underperforming, the WWTPs in the canton of Vaud are now undergoing a regionalisation process, which will regroup wastewater treatment so that the wastewater from several towns can be treated by a smaller number of new, more modern and effective WWTPs.

9.6.2 Nemiña beach (Spain)

9.6.2.1 General information

- Region: City of Muxia – Province of A Coruña – Galicia (Spain)
- Affected waters: the mouth of the Castro river; the spill appears to have been into the sea.
- Plants: Several municipal WWTPs discharge their effluent into the river. However, no biomedia were found upstream of the river mouth. A large fish farm (trout) is located at the mouth of the river, with some discharges being made directly into the sea.
- Type of biomedia found: K1

9.6.2.2 Account of the incident

In mid-November 2017, volunteers reported large numbers of biomedia on the beach at Nemiña in the town of Muxia (Spain). On 16 and 17 November 2017, they collected over 900 biomedia all of the same kind (K1), as well as a few samples of a different model. In the following weeks around 150 to 200 biomedia were collected from the same beach. On 4 January 2018, 288 biomedia were found in a 50-metre transect, with some 698 biomedia counted along the whole beach.

The absence of biomedia further upstream and the presence of these plastic pieces in such large quantities on the beach nearest to the river's mouth make it seem likely that a major spill happened at a facility in the immediate vicinity of the beach. The absence of biomedia on neighbouring beaches could be explained by the specific currents in the area and the orientation of the beach. The biomedia found looked new (they have not undergone any alteration due to lengthy presence in the marine environment) which made this seem like a recent spill.

9.6.2.3 Actions undertaken

Local volunteers alerted the Muxia town hall, the police and media. This action did not result in any response from the local authority, and the police and press did not show much interest either. Surfrider Foundation Europe has conducted investigations to find out if the companies located in the immediate proximity are using a wastewater treatment process, or if a local WWTP has experienced any incidents but no official proof has been found.

9.6.3 Evaluation of observed pollution events

Out of the fifteen major pollution incidents investigated by Surfrider Foundation, nine can be clearly linked to failures at WWTPs. All of the incidents at WWTPs leading to biomedia spilling out into the natural environment were the result of heavy rainfall, which caused blockages and even overflows, which were difficult to manage.

The vagaries of weather and lack of awareness of the potential impact of possible biomedia spills are some of the greatest problems at WWTPs. Out of the ten major pollution incidents reported in Europe, none of them resulted in any effective warning being issued by the WWTP managers, which in turn led to the biomedia spreading across huge distances in the environment.

9.7 SYSTEM MALFUNCTIONS

The various pollution incidents underscore just how vulnerable these installations are to weather events. On top of this, there are very few measures in place to raise the alarm in the event of incidents relating to the use of biomedia. Comparing the different incidents has enabled a list of the main kinds of malfunctions reported to be put together, making it easier to understand the causes of the problems.

9.7.1 Cause of system malfunctions

On-site investigations and a study of the literature on this subject show that the main reason for losses of biomedia into the environment is due to overflows of the tanks in which they are held. In order to work out the possible causes of these overflows, it is important to look at how bioreactors are configured in order to focus on the possible critical spill points.

Many bioreactors are not hermetically sealed. They have various influent and effluent channels to enable untreated water to enter the tank, chemical agents to be added to the tank to treat the water, treated water to leave the tank, and excess water to be removed from the reactor. If things go wrong and the water level in the reactor rises, any of these channels can provide a means for biomedia to spill out into the environment. In addition, because the reactors are not always covered, overflows can even occur over the edges of the reactor itself.

9.7.2 Description of observed cases

9.7.2.1 Blockage of bioreactor effluent mesh

Effluent mesh becoming blocked is the number one cause of system malfunctions detected and can have various causes. Blockages are caused by the biomedia obstructing the grilles covering the tank's exit points. The flow of water leaving the tank carries the plastic biomedia with it, causing them to get stuck against the mesh. This obstruction reduces the flow of water leaving the tank, creating a differential between the influent and effluent flow, and causing the water level in the tank to rise until it overflows.

Various potential causes of blockages have been identified:

- the WWTP was not adapted for use with biomedia: biomedia have been added to the reactors to boost their treatment capacity but the flat effluent mesh has not been replaced by cylindrical mesh;

- the biomedia are not being agitated: lack or failure of aeration systems, mixing rotors or flow inversion systems;
- poor process management strategy: the plant manager decides to reduce biomedia agitation to a level below the manufacturer's guidelines in order to save energy;
- sensor malfunction: the sensors in the reactor are used to test oxygen levels. When this level becomes too high, the sensors reduce the aeration levels and agitation of the biomedia. This can occur when storms result in large volumes of oxygen-rich water entering the system;
- selection of a biofilm carrier that is not suitable for the intended use: some biomedia have a propensity to stick together and form clumps if they are not mixed with other models that allow them to detach from each other.

9.7.2.2 Excess aeration

Excessive aeration levels in the basins due to poor system settings, human error or exceptional weather can cause the plastic biomedia to trap air bubbles in their cavities. This dramatically decreases their density, so they float to the surface and can potentially escape by overflowing the tank if the water level is high.

9.7.2.3 Failure of safety systems

Sensors situated at different key points around the wastewater treatment system measure flows and open secondary channels in the event of any problem. However, faults with these can lead to overflows and loss of biomedia (Figure 6.1).

9.7.2.4 Commissioning of a new WWTP

Problems can occur when a new WWTP is commissioned. Theoretical calculations can be quite different from actual conditions on the ground or from the reality of the completed project, and this can lead to losses.

9.7.2.5 Limitations of the combined sewer system

In many localities, wastewater is still collected in a combined system. During periods of heavy rain WWTPs can receive excessive amounts of water, leading to overflows from the treatment tanks and losses of biomedia into the environment.

9.7.2.6 Poor storage of biomedia

The way in which biomedia are stored can result in losses even before a biomedia process is put into operation at a plant. Biomedia can spill from sacks during handling, and if these sacks are stored unprotected and in the open, this can also lead to spills or overflows during extreme weather events (rain and wind).

9.7.2.7 Diffuse pollution

Some biomedia models can be found along river banks and coasts all year round in small quantities. These could be the remnants of major spills still turning up years later but could also be from one-off small losses.

9.8 CONCLUSIONS

Biomedia can be found wherever water needs to be treated, be this in municipal or industrial WWTPs, at vineyards, fish farms, livestock farms or even in private dwellings. We have been able to better understand the source of pollution incidents since 2010 thanks to the numerous finds of biomedia along our coasts, and due to eyewitness reports, interviews and the heavy involvement of numerous volunteers.

Given the widespread dispersal of this type of pollution once in the environment, and particularly in the sea, its origin is often difficult to trace. This is why it is essential to act at the source of any potential pollution, starting from the use of the biomedia. A good understanding of the environmental risks associated with the use of biomedia, from the earliest stages of setting up WWTPs, is critical. Above all, this involves raising awareness, particularly among WWTP operators, who should in any case not ignore the impact of biomedia pollution.

Our study has revealed a lack of reactiveness and responsibility on the part of WWTP operators when incidents occur leading to spills of biomedia. At a European level, clean-up actions following pollution events are the exception rather than the rule. This means biomedia from spills that took place over five years ago can still be found polluting the environment and coasts. Biomedia lost regularly but in small quantities into watercourses also contribute to this source of permanent pollution.

The use of biomedia in wastewater treatment processes is growing exponentially, which in turn increases the risk of incidents. This is why it is so important to implement information and prevention measures, and protocols for raising the alarm, as well as additional low-cost steps that could help to largely prevent biomedia losses and reduce the risk of pollution if they get out into the environment.

REFERENCES

Araújo M. C. and Costa M. F. (2007). An analysis of the riverine contribution to the solid wastes contamination of an isolated beach at the Brazilian Northeast. *Management of Environmental Quality*, **18**, 6–12.

Bencivengo P., Barreau C., Bailly C. and Verdet F. (2018). Sewage Filter Media and Pollution of the Aquatic Environment, Surfrider Foundation Europe Report. Water Quality and Marine Litter programme, Biarritz, France. https://www.surfrider.eu/wp-content/uploads/2018/08/biomedia-pollution-report.zip (accessed August 2018).

Canler J. P. and Perret J. M. (2012). Les procédés MBBR pour le traitement des eaux usées, cas du procédé R3F (MBBR systems for wastewater treatment, case of the R3F process). *FNDAE Technical Document*, **38**. FNDAE, France.

Canler J. P., Perret J. M. and Choubert J. M. (2012). Évaluation, optimisation et modélisation de filières de traitement: cas du procédé à cultures fixées fluidisées (MBBR) (Evaluation, optimization and modeling of treatment systems: the case of the MBBR). *Sciences Eaux et Territoires: la Revue du IRSTEA (Sciences, Water & Territories, IRTSEA's Journal)*, **9**, 16–23.

Gonzalez G., Herrera M. G., Garcia M. T. and Pena M. M. (2001). Biodegradation of phenol in a continuous process: comparative study of stirred tank and fluidized-bed bioreactors. *Bioresource Technology*, **76**, 245–251.

Jambeck J. R., Geyer R., Wilcox C., Siegler T. R., Perryman M, Andrady A., Narayan R. and Lavander Law K. (2015). Plastic waste inputs from land to the ocean. *Science*, **347** (6223), 768–771.

Jianping W., Lei P., Lipping D. and Guozhu M. (2003). The denitrification treatment of low C/N2 ratio nitrate-nitrogen wastewater in a gas–liquid–solid fluidized bed bioreactor. *Chemical Engineering Journal*, **94**, 155–159.

Kargi F. and Karapinar I. (1997). Performance of fluidized bed bioreactor containing wire-mesh sponge particles in wastewater treatment. *Waste Management*, **17**(1), 65–70.

Laurent J. (2006). Etude du fonctionnement d'un réacteur à lit fluidisé et à alimentation séquentielle (Study of the functioning of a fluidized bed reactor with a sequential supply). Rapport de recherche pour l'obtention du master recherche chimie et microbiologie de l'eau – Université de Limoges. (Report for Master's Degree in Chemistry and Microbiology of Water, University of Limoges.)

Mazioti A. A., Stasinakis A. S., Pantazi Y. and Andersen H. R. (2015). Biodegradation of benzotriazoles and hydroxy-benzothiazole in wastewater by activated sludge and moving bed biofilm reactor systems. *Bioresource Technology*, **192**, 627–635.

Nicolella C., Van Loosdrecht M. and Heijnen J. (2000). Wastewater treatment with particulate biofilm reactors. *Journal of Biotechnology*, **80**, 1–33.

Venu Vinod A. and Venkat Reddy G. (2005). Simulation of biodegradation process of phenolic wastewater at higher concentrations in a fluidized-bed bioreactor. *Biochemical Engineering Journal*, **24**, 1–10. https://doi.org/10.1016/j.bej.2005.01.005.

Chapter 10

Effects of microplastics on freshwater and marine microalgae

Merve Tunalı and Orhan Yenigün

Boğaziçi University, Institute of Environmental Sciences, Bebek, Istanbul, Turkey

Keywords: Algal growth, Chlorophyll concentration, Combined effects, Photosynthesis, Surface charge

10.1 THE GLOBAL PLASTIC PROBLEM

Plastics are now one of the essential materials widely used in daily life for consumer products as well as in industrial processes. Being light, durable, low cost, resistant to most chemicals and easy to process (Li *et al.*, 2016; Thompson *et al.*, 2009), they have been used in various applications since the early 1900s (Wong *et al.*, 2015). Usage of plastics around the world has increased dramatically, especially in the last few decades, with 47 million tons, 288 million tons and 335 million tons produced in 1976, 2002 and 2015, respectively (Plastics Europe, 2013, 2017), and production is expected to double again in the next 20 years (Lyakurwa, 2017).

Increasing plastic production results in increasing amounts of plastic waste in terrestrial and marine ecosystems, and plastic debris can be found around the world from the Arctic to the Mediterranean and the Pacific Ocean (Lagarde *et al.*, 2016). It is estimated that, annually, 8 million tons of plastics are released to oceans (Lackey, 2018). Marine-based plastics mostly come from land-based sources; 80% of marine plastic debris is derived from sources such as coastal recreational activities, wastewater effluents and leachates, and from solid waste

Editors: Hrissi K. Karapanagioti and Ioannis K. Kalavrouziotis
doi: 10.2166/9781789061697_0147

disposal. They are transported by rivers, streams and wastewater treatment systems into the marine environment. The rest comes from ocean-based activities such as commercial fishing (Li *et al.*, 2016). Plastics are also found in other water bodies, such as rivers and lakes (Auta *et al.*, 2017; Lagarde *et al.*, 2016; Li *et al.*, 2016; Tang *et al.*, 2018; Yokota *et al.*, 2017). Being non-biodegradable, they may stay in ecosystems for many years (Cole *et al.*, 2011), effecting organisms and primary producers, as well as resulting in an accumulation problem.

Large plastic debris (referred to as "macroplastics") are a matter of concern for their effects on water bodies over a long period of time, and cause aesthetic problems as well as threatening sailing, fishing and aquaculture. Further, they cause injury and death to marine organisms, negatively affect gas exchange between pore waters and overlying sea water and may create artificial seabeds (Cole *et al.*, 2011; Gregory, 2009).

There has also been an increasing concern about microplastics – tiny plastic fragments, fibers and granules – over the past decade. Microplastics are categorized as small pieces of plastic, generally defined as <5 mm in diameter (GESAMP, 2015; Isensee & Valdes, 2015) resulting from the break down of bigger particles (Andrady, 2011; Vince & Stoett, 2018). As bigger chunks are broken down into tiny pieces, an immense increase in the plastic's surface area is observed, which enables the microplastics to get into contact with nearly everything in the marine environment. Other sources of microplastics include microfibers, marine paints, the sandblasting industry (Niaounakis, 2017), cosmetics and personal hygiene products (Auta *et al.*, 2017; Niaounakis, 2017). Commonly found microplastic types include polyvinyl chloride (PVC), nylons and polyethylene (PE), polypropylene (PP), polystyrene (PS), and polyethylene terephthalate (PET), representing 95% of the world plastics production (Andrady, 2011; Brien, 2007; Lagarde *et al.*, 2016).

These microplastics have negative effects on ecosystems by entering the food web and acting as a vector for toxic chemical transfer (Cole *et al.*, 2011; Duncan *et al.*, 2018). Many marine species are known to be affected by plastic debris (Li *et al.*, 2016; Lusher, 2015; Tang *et al.*, 2018; Zhang *et al.*, 2017) and the effects of microplastics on primary producers (the basis of the food chain) are also now in focus (MacPhee, n.d.; Sjollema *et al.*, 2016). Among these primary producers, microalgae have a crucial role in maintaining the ecosystem balance (Harris, 1986; Zhang *et al.*, 2017) since much of the oxygen in the world is supplied by algal photosynthesis (Bhattacharya *et al.*, 2010).

10.2 EFFECTS OF MICROPLASTICS ON MICROALGAE

Details of studies on the effects of plastics on microalgae can be seen in Table 10.1.

10.2.1 Algal growth

The factors effecting algal growth are described below.

Table 10.1 Effects of plastics on microalgae.

Category	Effect	Plastic Type	Plastic Size	Concentration	*Algae Type*	Exposure Time	Reference
Algae growth	Decreased by 45% (for 0.05 µm in 250 mg/L); Decreased by 11% (for 0.5 µm in 250 mg/L)	Polystyrene	0.05 µm, 0.5 µm and 6 µm	25, 250 mg/L	*Dunaliella tertiolecta*	72 h	Sjollema *et al.* (2016)
	Decreased by 39.7% (in 50 mg/L) (Decreased with increasing concentration)	Polyvinyl chloride	1 µm	0, 1, 5, 10, 50 mg/L	*Skeletonema costatum*	96 h	Zhang *et al.* (2017)
	No effect	Polyvinyl chloride	1 mm	0, 50, 500, 1,000, 2,000 mg/L	*Skeletonema costatum*	96 h	Zhang *et al.* (2017)
	No effect (in 75 particles/mL) Decreased (in 7,500 particles/mL)	Polystyrene	10 µm	75, 750 and 7,500 particles/mL	*Rhodomonas baltica.*	264 h	Lyakurwa (2017)
	Decreased (after 41.5 g/L)	Red fluorescent polymer microspheres	1 to 5 µm	0.75, 1.5, 3, 6, 12, 24, 48 mg/L	*Tetraselmis chuii*	96 h	Prata *et al.* (2018)
	Decreased by 18% (for polypropylene, only after 78 days) No effect for high density polyethylene	Polypropylene and high density polyethylene	400 to 1,000 µm	400 mg/L	*Chlaydoas reinhardtii*	>78 d	Lagarde *et al.* (2016)
	Decreased by 24% (in 1.472 mg/L)	Polyethylene	1 to 5 µm	0.046, 0.092, 0.184, 0.368, 0.736, 1.472 mg/L	*Tetraselmis chuii*	96 h	Davarpanah and Guilhermino (2015)
	No effect	Polystyrene	2 µm	3.96 mg/L	*Tisochrysis lutea, and Chaetoceros neogracile*	>30 d	Long *et al.* (2017)
	Decreased by 2.5% (in 1 g/L PS)	Polystyrene	70 nm	44 to 1,100 mg/L	*Scenedesmus obliquues*	72 h	Besseling *et al.* (2014)
Chlorophyll concentration	Decreased by 5% (in 5 mg/L) Decreased by 32% (in 50 mg/L)	Polyvinyl chloride	1 µm	5 to 50 mg/L	*Skeletonema costatum*	96 h	Zhang *et al.* (2017)
	Decreased chlorophyll concentration under the exposure of high microplastic concentration (7,500 particles/mL) After algae reaches stationary phase, all samples showed decreased chlorophyll concentration.	Polystyrene	10 µm	75, 750 and 7,500 particles/mL	*Rhodomonas baltica.*	264 h	Lyakurwa (2017)

(*Continued*)

Table 10.1 Effects of plastics on microalgae. (*Continued*)

Category	Effect	Plastic Type	Plastic Size	Concentration	*Algae Type*	Exposure Time	Reference
	No effect	Polystyrene	2 μm	3.96 mg/L	*Tisochrysis lutea, and Chaetoceros neogracile*	> 30 d	Long *et al.* (2017)
	Decreased by 46% (in 0.9 mg/L) Decreased by 37% (in 2.1 mg/L)	Red fluorescent polymer microspheres	1 to 5 μm	0.75, 1.5, 3, 6, 12, 24, 48 mg/L	*Tetraselmis chuii*	96 h	Prata *et al.* (2018)
Photosynthetic efficiency	Negligible effect (<10% decrease)	Polystyrene	0.05 μm, 0.5 μm and 6 μm	25, 250 mg/L	*Dunaliella tertiolecta*	72 h	Sjollema *et al.* (2016)
	Decreased	Polystyrene	20 nm	1.6–40 mg/mL	*Chlorella and Scenedesmus*	>70 h	Bhattacharya *et al.* (2010)
Combined effect (with pharmaceuticals)	Decreased growth rate Decreased chlorophyll content	Microplastics-procainamide and microplastics-doxycycline mixtures	1–5 μm	Microplastics-procianamice mixture; procainamide: between 8 to 256 mg/L with 1.5 mg/L microplastics, Microplastics-doxycycline mixture:doxycycline between 4 to 128 mg/L with 1.5 mg/L microplastics	*Tetraselmis chuii*	96 h	Prata *et al.* (2018)
Combined effect (with copper)	No effect	Polyethylene	1–5 μm	Copper between 0.02 to 0.64 mg/L in the presence of 0.184 mg/L microplastic	*Tetraselmis chuii*	96 h	Davarpanah and Guilhermino (2015)
Adsorption of plastics to microalgae	Increased in positively charged MP, Negligible effect in negativly charged	Polystyrene	20 nm	1.6 to 40 mg/mL	*Chlorella and Scenedesmus*	>70 h	Bhattacharya *et al.* (2010)
Uptake of microplastics by algae	Ingested by algae	Polystyrene, green fluorescent	10 μm virgin polystyrene, 1–5 μm green fluorescent	75 and 7,500 microbeads/mL	*Oxyrrhis marina*	1 h	Lyakurwa (2017)

10.2.1.1 Concentration

Microplastics generally affect algal growth negatively, depending on concentration and size of the microplastics. Many studies have shown that inhibition of algal growth occurs above a defined microplastic concentration. Prata *et al.* (2018) tested the effect of red fluorescent polymer microspheres on *Tetraselmis chuii*, a type of marine microalgae, studying a concentration range from 0.75 to 48 mg/L. Decrease in algal growth was found above 41.5 mg/L microplastic concentration. Similar results were found by Besseling *et al.* (2014) and Lyakurwa (2017), who tested polystyrene particles on *Rhodomonas baltica*; it was found that 75 particles/mL had no effect on algal growth, while 7500 particles/mL reduced growth (Lyakurwa, 2017). The inhibitory effect of polystyrene particles on *Scenedesmus obliquues* increased with increasing concentration; tested concentrations were from 44 to 1100 mg/L and growth inhibition reached 2.5% in the presence of 1 g/L microplastics (Besseling *et al.*, 2014). Davarpanah and Guilhermino (2015) tested the effect of changes of polyethylene concentration (from 0.046 to 1.472 mg/L) on the marine microalgae, *Tetraselmis chuii.* As the concentration of the microplastic increased, there was a measurable decrease in the specific growth rate of the algae, reaching a maximum 24% growth inhibition. However, inhibition did not significantly increase with increased concentration. This was due to the aggregation and sedimentation of the plastics (Luís *et al.*, 2015) or to the relatively low concentration of plastics in the defined study. The magnitude of the inhibitory effect varies between studies which may be because of the type of algae studied. In some studies, no measurable effect on algal growth was observed; polystyrene with a concentration of 3.96 mg/L did not effect the growth of *Tisochrysis lutea* and *Chaetoceros neogracile* (Long *et al.*, 2017), with the authors concluding that the concentration was too low to affect the growth of these species.

In general, the reason for the inhibitory effect on algal growth may be because of the microplastics' limiting effect on the energy and substance transfer between the cells and the environment. Thus, a decrease in nutrient intake, carbon dioxide and oxygen transport as well as light from media into cells may be observed (Zhang *et al.*, 2017). It has also been found that the decrease in growth is not related to the shading effect of microplastics (Sjollema *et al.*, 2016; Zhang *et al.*, 2017). Another reason may be that harmful metabolites may not be able to exit cells and thus limit their growth (Zhang *et al.*, 2017). Reduced energy and a toxic effect on growth may also be caused by biofilm formation, where microalgae accumulate on the surface of microplastics (Andrady, 2011; Lyakurwa, 2017). Aggregation may also occur (Lagarde *et al.*, 2016; Ma *et al.*, 2014) and this may cause precipitation.

10.2.1.2 Size of the microplastics

Size also has significant effects on growth rate. Smaller sized particles have much more of an effect on organisms at a cellular level since, as their size decreases, the possibility of passing through the cell membrane increases (Lusher, 2015). Sjollema *et al.* (2016) studied the effect of polystyrene on microalgae, using three different sizes of plastics (0.05, 0.5 and 6 μm) in 25 and 250 mg/L solutions. They found that only the small particles (0.05 μm) at higher concentrations affected the growth of *Dunaliella tertiolecta*. The growth of the algae was found to be reduced by 45% in the presence of 0.05 μm, while 11% growth reduction was observed for the same algae for the 0.5 μm sized plastics. Similar results were found by Zhang *et al.* (2017) who investigated marine microalgae (*Skeletonema costatum*) for the toxic effects of polyvinyl chloride microplastics (mPVC) with average diameter of 1 μm and polyvinyl chloride bulk plastic debris (bPVC) with average diameter of 1 mm. The algal density was reduced by 39.7% in 50 mg/L mPVC, while bPVC had no effect on algal growth. In the same study it was shown that the growth reduction was greater with increasing concentration in the case of mPVC. This may be because the bulk particles float and do not have a chance to interact with the microalgae (Zhang *et al.*, 2017).

10.2.1.3 Effect of time

Zhang *et al.* (2017) found that time had no measurable effect on *Skeletonema costatum* when exposed to polyvinyl chloride. However, another study (Lagarde *et al.*, 2016) showed that growth suppression may be dependent on time in an investigation of the effect of polypropylene and high density polyethylene on freshwater microalgae (*Chlaydoas reinhardtii*). The size range of the plastics was 400–1000 μm and it was found that polypropylene and high density polyethylene particles had no influence on algal growth until day 63. However, a measurable decrease in growth was observed (18%) after 78 days of polypropylene exposure. The duration of the two studies, the microplastics and algal types, as well as the different concentrations and sizes of microplastics used are all shown in Table 10.1.

10.2.2 Photosynthetic efficiency and chlorophyll concentration

Another effect of microplastics on microalgae is a decrease in chlorophyll concentration and photosynthetic activity. Zhang *et al.* (2017) investigated the negative effects of polyvinyl chloride microplastics on chlorophyll concentration and photosynthetic efficiency, finding that chlorophyll concentration decreased by 7% and 20% in 5 and 50 mg/L solutions, respectively, and that photosynthetic efficiency decreased by 5% and 32% in 5 and 50 mg/L, respectively. Lyakurwa (2017) showed that chlorophyll production also decreased with increased plastic concentration until a stationary phase. The results of the study conducted by Bhattacharya *et al.* (2010) is also in line with these results, finding that respiration occurs faster that photosynthesis resulting in an extended effort to gain motility.

Besseling *et al.* (2014) found that the effect on chlorophyll concentration is significant above 100 mg/L for 70 nm sized microplastics, while Prata *et al.* (2018) showed that there is a decrease in chlorophyll concentration in the presence of 0.9 mg/L microplastic consisting of 1–5 μm sized particles; no concentration–response relationship was observed. Sjollema *et al.* (2016) showed that there is negligible effect on photosynthetic efficiency in the presence of 250 mg/L of 0.05, 0.5 and 6 μm sized particles. No effect was found by Long *et al.* (2017) in the presence of 3.96 mg/L of 2 μm sized polystyrene. Therefore; microplastic waste can have various impacts depending on the algal species present in the water body and concentration of the contaminant. The negative effect on photosynthesis may be because of physical toxicity; however, the mechanism of toxicity is not yet clearly known (SAPEA, 2019).

10.2.3 Other effects

Other effects of microplastics on microalgae are described below.

10.2.3.1 Surface charge

Surface charge is another factor that may have an influence on the effect of microplastics on microalgae (Auta *et al.*, 2017). Charge affects particle stability (Alimi *et al.*, 2018) and microplastic attachment (Yokota *et al.*, 2017). Bhattacharya *et al.* (2010) showed that adsorption of positively charged microplastics on algal species caused more production of reactive oxygen species than it did of negative ones. However, Sjollema *et al.* (2016) showed that negatively charged microplastics have no effect on the growth of *D. Tertiolecta* compared to uncharged ones. The effect of polystyrene was assessed on *Chlorella* and *Scenedesmus* by Bhattacharya *et al.* (2010). Increase in adsorption with increased concentration of polystyrene was observed for positively charged microplastics. However, in negatively charged polysytrene, the adsorption rate is very low. When the type of the algae was considered, positively charged microplastics affinity was found to be higher in *Scenedesmus*, while it was lower in negatively charged ones. This may be because of differences in morphology and because of the difference in total surface area (Bhattacharya *et al.*, 2010).

10.2.3.2 Combined effects

Chemicals coming from their manufacturing may also be present on microplastics (GESAMP, 2015; SAPEA, 2019) and other substances present in water bodies may interact with microplastics. So, there may be combined effects that should also be considered. Prata *et al.* (2018) tested the combined effect of microplastics with pharmaceuticals. When the effect of microplastic-procainamide and microplastic-doxycycline mixtures were tested, it was found that the mixtures had a more toxic effect than each of the substances tested alone. This may be because of microplastic interaction with the cell wall that supports the uptake of pharmaceutical substances. Okubo *et al.* (2018) recently studied the effects of microplastics on the initiation of symbiotic relationships in anthozoan-algae symbiosis and found that microplastics damage the symbiotic relationship. When the combined effect of copper and microplastics was tested, no significant differences were found in the presence of 0.02–0.64 mg/L copper together with 1 to 5 μm sized 0.184 mg/L microplastics (Davarpanah & Guilhermino, 2015). This may be because of the relatively low concentration of copper and microplastics tested.

10.2.3.3 Others

Casabianca *et al.* (2018) analysed samples of marine plastics in terms of the attachment of harmful microalgae, finding that the plastics provide a substrate that harmful microalgae can attach to, colonize rapidly and produce toxins. Lyakurwa (2017) used marine microalgae; *Oxyrrhis marina*, to test the ingestion of microplastics finding that the microplastics were ingested by *Oxyrrhis marina*. As the concentration of the microplastics increased, loss of motility and food replacement was observed. Another effect found in the literature is hetero-aggregate formation. Lagarde *et al.* (2016) demonstrated aggregation in the presence of polypropylene in 20 days; however, no aggregation occurred in the presence of high density polyethylene particles.

10.3 CONCLUSIONS

The amount of microplastics found in freshwater and marine ecosystems are increasing significantly, related to the increasing production and consumption of plastics. Microplastics have negative effects on organisms and primary producers in the systems in which they end up. The main effects of microplastics on microalgae are decreased algal growth, chlorophyll concentration and photosynthetic efficiency. These effects have been found to be dependent on type, size and concentration of the plastics and algal type. Other studied effects include combined effects with other substances such as copper and pharmaceuticals. Combined exposure with microplastics may increase the effects on microalgae; depending on the type of material. In addition, surface charge also has an influence on effect. Other effects found in the literature include microplastics acting as a substrate for algal growth, ingestion of microplastics by microalgae and hetero-aggregation. Even though many other effects, especially combined effects, are not widely studied yet, it can be said that, as microplastic pollution increases, the impact of its effects on the ecosystem seems to become more problematic starting from the bottom of the food chain.

It should be noted that the types of plastics and types of algae tested should be extended in order to fill the gaps in the literature. In addition, more focus should be given to combined effects of microplastics with other materials to stimulate the natural environment and to quickly understand the complexity of the problem.

REFERENCES

Alimi O. S., Farner Budarz J., Hernandez L. M. and Tufenkji N. (2018). Microplastics and nanoplastics in aquatic environments: aggregation, deposition, and enhanced contaminant transport. *Environmental Science & Technology*, **52**(4), 1704–1724.

Andrady A. L. (2011). Microplastics in the marine environment. *Marine Pollution Bulletin*, **62**(8), 1596–1605. doi:10.1016/j.marpolbul.2011.05.030.

Auta H. S., Emenike C. and Fauziah S. (2017). Distribution and importance of microplastics in the marine environment: a review of the sources, fate, effects, and potential solutions. *Environment International*, **102**, 165–176.

Besseling E., Wang B., Lürling M. and Koelmans A. A. (2014). Nanoplastic affects growth of *S. obliquus* and reproduction of *D. magna*. *Environmental Science & Technology*, **48** (20), 12336–12343. doi: 10.1021/es503001d.

Bhattacharya P., Lin S., Turner J. P. and Ke P. C. (2010). Physical adsorption of charged plastic nanoparticles affects algal photosynthesis. *Journal of Physical Chemistry C*, **114**(39), 16556–16561. doi: 10.1021/jp1054759.

Brien S. (2007). Vinyls Industry Update. Presentation at the World Vinyl Forum2007. See: http://vinyl-institute.com/Publication/WorldVinylForumIII/VinylIndustryUpdate.aspx (accessed October 2018).

Casabianca S., Capellacci S., Giacobbe M. G., Dell'Aversano C., Tartaglione L., Varriale F., Narizzano R., Risso F., Moretto P., Dagnino A., Bertolotto R., Barbone E., Ungaro N. and Penna A. (2018). Plastic-associated harmful microalgal assemblages in marine environment, *Environmental Pollution*, **244**, 617–626.

Cole M., Lindeque P., Halsband C. and Galloway T. S. (2011). Microplastics as contaminants in the marine environment: a review. *Marine Pollution Bulletin*, **62**(12), 2588–2597. doi: 10.1016/j.marpolbul.2011.09.025.

Davarpanah E. and Guilhermino L. (2015). Single and combined effects of microplastics and copper on the population growth of the marine microalgae *Tetraselmis chuii*. Estuarine, *Coastal and Shelf Science*, **167**, 269–275.

Duncan E. M., Arrowsmith J., Bain C., Broderick A. C., Lee J., Metcalfe K. and Godley B. J. (2018). The true depth of the Mediterranean plastic problem: extreme microplastic pollution on marine turtle nesting beaches in Cyprus. *Marine Pollution Bulletin*, **136**, 334–340.

GESAMP (2015). Chapter 3.1.2 Defining 'microplastics'. Sources, fate and effects of microplastics in the marine environment: a global assessment. In: IMO/FAO/UNESCO–IOC/UNIDO/WMO/IAEA/UN/UNEP/UNDP Joint Group of Experts on the Scientific Aspects of Marine Environmental Protection (GESAMP), P. J. Kershaw, (ed.), Rep. Stud. GESAMP No. 90, 96p, London.

Gregory M. R. (2009). Environmental implications of plastic debris in marine settings: entanglement, ingestion, smothering, hangers-on, hitch-hiking and alien invasions. *Philosophical Transactions of the Royal Society B: Biological Sciences*, **364**, 2013–2025.

Harris G. P. (1986). Phytoplankton Ecology: Structure, Function and Fluctuation. Chapman and Hall Ltd., London.

Isensee K. and Valdes L. (2015). Marine Litter: Microplastics. GSDR 2015 Brief. See: https://sustainabledevelopment.un.org/content/documents/5854Marine%%20-%20Microplastics.pdf (accessed October 2018).

Lackey K. (2018). From trash to treasure: Ocean plastic waste source of alternative energy. *World Water*, **41**, 14–16.

Lagarde F., Olivier O., Zanella M., Daniel P., Hiard S. and Caruso A. (2016). Microplastic interactions with freshwater microalgae: hetero-aggregation and changes in plastic density appear strongly dependent on polymer type. *Environmental Pollution*, **215**, 331–339. doi: 10.1016/j.envpol.2016.05.006.

Li W. C., Tse H. F. and Fok L. (2016). Plastic waste in the marine environment: a review of sources, occurrence and effects. *Science of the Total Environment*, **566–567**, 333–349.

Long M., Paul-Pont I., Hégaret H., Moriceau B., Lambert C., Huvet A. and Soudant P. (2017). Interactions between polystyrene microplastics and marine phytoplankton lead to species-specific hetero-aggregation. *Environmental Pollution*, **228**, 454–463. doi: 10.1016/j.envpol.2017.05.047.

Luís L. G., Ferreira P., Fonte E., Oliveira M. and Guilhermino L. (2015). Does the presence of microplastics influence the acute toxicity of chromium (VI) to early juveniles of the common goby (*Pomatoschistus microps*)? A study with juveniles from two wild estuarine populations. *Aquatic Toxicology*, **164**, 163–174.

Lusher A. (2015). Microplastics in the marine environment: distribution, interactions and effects. In: Marine Anthropogenic Litter, M. Bergmann, L. Gutow and M. Klages (eds.). Springer, Cham.

Lyakurwa D. J. (2017). Uptake and effects of microplastic particles in selected marine microalgae species; *Oxyrrhis marina* and *Rhodomonas baltica*. Master of Science thesis, Norwegian University of Science and Technology Department of Biology, Norway.

Ma S., Zhou K., Yang K. and Lin D. (2014). Heteroagglomeration of oxide nanoparticles with algal cells: effects of particle type, ionic strength and pH. *Environmental Science & Technology* **49**, 932–939.

MacPhee L. (n.d.) *Life on the Food Chain*. Northern Arizona University. See: https://www2.nau.edu/lrm22/lessons/food_chain/food_chain.html (accessed October 2018).

Niaounakis M. (2017). The problem of marine plastic debris. In: Management of Marine Plastic Debris, Elsevier Inc. pp. 1–55.

Okubo N., Takahashi S. and Nakano Y. (2018). Microplastics disturb the anthozoan-algae symbiotic relationship. *Marine Pollution Bulletin*, **135**, 83–89. doi: 10.1016/j.marpolbul.2018.07.016.

Plastics Europe (2013). Plastics – the Facts 2013: An Analysis of European Latest Plastics Production, Demand and Waste Data. Association of Plastics Manufacturers, Belgium

Plastics Europe (2017). Plastics – the Facts 2017. An Analysis of European Plastics Production, Demand and Waste Data. Association of Plastics Manufacturers, Belgium.

Prata J. C., Lavorante B. R. B. O., B. S. M. Montenegro M. da C. and Guilhermino L. (2018). Influence of microplastics on the toxicity of the pharmaceuticals procainamide and doxycycline on the marine microalgae *Tetraselmis chuii*. *Aquatic Toxicology*, **197**, 143–152. doi: 10.1016/j.aquatox.2018.02.015.

Science Advice for Policy by European Academies (SAPEA) (2019). A Scientific Perspective on Microplastics in Nature and Society. SAPEA, Berlin. https://doi.org/10.26356/microplastics

Sjollema S. B., Redondo-Hasselerharm P., Leslie H. A., Kraak M. H. S. and Vethaak A. D. (2016). Do plastic particles affect microalgal photosynthesis and growth? *Aquatic Toxicology*, **170**, 259–561.

Tang J., Ni X., Zhou Z., Wang L. and Lin S. (2018). Acute microplastic exposure raises stress response and suppresses detoxification and immune capacities in the scleractinian coral *Pocillopora damicornis*. *Environmental Pollution*, **243**, 66–74.

Thompson R. C., Moore C. J., vom Saal F. S. and Swan S. H. (2009). Plastics, the environment and human health: current consensus and future trends. *Philosophical Transactions of the Royal Society B: Biological Sciences*, **364**(1526), 2153–2166.

Vince J. and Stoett P. (2018). From problem to crisis to interdisciplinary solutions: plastic marine debris. *Marine Policy*, **96**, 200–203.

Wong S. L., Ngadi N., Abdullah T. A. T. and Inuwa I. M. (2015). Current state and future prospects of plastic waste as source of fuel: a review. *Renewable and Sustainable Energy Reviews*, **50**, 1167–1180. doi:10.1016/j.rser.2015.04.063.

Yokota K., Waterfield H., Hasting C., Davidson E., Kwietniewski E. and Wells B. (2017). Finding the missing piece of the aquatic plastic pollution puzzle: interaction between primary producers and microplastics. *Limnology and Oceanography Letters* **2**, 91–104.

Zhang C., Chen X., Wang J. and Tan L. (2017). Toxic effects of microplastic on marine microalgae Skeletonema costatum: interactions between microplastic and algae. *Environmental Pollution*, **220**, 1282–1288. doi: 10.1016/j.envpol.2016.11.005.

Chapter 11

Possible effects on plants due to microplastics in soils from wastewater effluent reuse or sewage sludge application

D. Papaioannou and I. K. Kalavrouziotis

Hellenic Open University, School of Science and Technology, Patra, Greece

Keywords: Agroecosystem, Biodegradation, Contaminant transport, Pollution

11.1 INTRODUCTION

Modern society and especially large urban populations are characterized by the production of high quantities of various industrial and personal by-products. Among these, in many developed regions, treated wastewater and sewage sludge are particularly interesting due to their agricultural application which has been evaluated as the most convenient recycling option for environmental and economic reasons (Kalavrouziotis & Koukoulakis, 2011).

Sewage sludge and wastewater have long been used as fertilizers and for crop irrigation, respectively. They have been used particularly in agriculture as a soil fertilizer (containing organic matter) and as supplementary sources of nitrogen and other macro- and micro-nutrients, as well as for improving soil physical conditions and productivity (Kalavrouziotis & Koukoulakis, 2011). However, in addition to significant soil-improving characteristics, sludge and wastewater carry a quite significant load of pollutants, such as heavy metals, toxic compounds, pharmaceutical xenobiotics and microplastics, and their long-term reuse may contribute to the accumulation of these pollutants in the soil (Papaioannou *et al.*, 2017).

Editors: Hrissi K. Karapanagioti and Ioannis K. Kalavrouziotis
doi: 10.2166/9781789061697_0159

In the last decade microplastics have become considered as a global environmental problem. Microplastic pollution has many potential sources such as industry, agriculture, landfill, wastewater treatment plants (WWTPs), household waste, tyres, synthetic textiles and packaging materials. Pollution from plastic materials (and its effects on marine life) was first recognized in marine environments in the early 1970s (Duis & Coors, 2016). Browne *et al.* (2011) were the first to point to and measure WWTP effluents as a source of microplastics. Despite the high rates of microplastics removal from WWTPs, even the small proportion of microplastics being released can result in significant amounts of microplastics entering the environment. Large plastic items which are present in water and soil gradually become smaller pieces because of various environmental weathering processes such as mechanical breakdown, decomposition and photo-degradation (Watts *et al.*, 2014). Knowledge is limited about the ecological impacts on the terrestrial environment – and especially on agro-ecosystems – by plastic pollution from wastewater and sewage sludge application and the potential consequences of microplastics in agriculture, including on sustainability and food safety (Bläsing & Amelung, 2018).

Microplastics are defined as plastic items which measure less than 5 mm in their longest dimension (GESAMP, 2015). Obviously, plastic particles in the environment continue to degrade and gradually become smaller until finally forming nanoplastics (Horton *et al.*, 2017), which are particles less than 100 nanometres (nm) in their longest dimension (Rios *et al.*, 2018).

Microplastics are emerging as anthropogenic pollutants of global importance. The environmental fate of microplastics in agro-ecosystems depends on complex interactions between soil physico-chemical, biological factors and the microplastics' properties. Consequently, it is important to study the sources of microplastics in agro-ecosystems, the mechanisms and behaviour of microplastics in soil and their reaction with soil organisms and plants.

11.2 MICROPLASTICS AND NANOPLASTICS IN AGRICULTURAL SOILS

11.2.1 Sources of plastic in agricultural soils

In developed regions, municipal effluents and urban runoff are eventually conveyed to WWTPs. In Europe, more than 45,000 WWTP units exist of which approximately 25,000 are primary and secondary, and about 20,000 are tertiary treatment plants (Amec Foster Wheeler, 2017). Generally, 1,270–2,130 tonne (t) microplastics per million inhabitants are released to urban environments, annually (Nizzetto *et al.*, 2016). Despite the satisfactory rates of removal of microplastics in wastewater treatment plants, a small quantity remains and leaves with the wastewater and sewage sludge, leading significant amounts of microplastics to enter the environment.

Generally, due to the low density of microplastics in WWTPs and the short time they spend there, degradation processes are still unknown. Studies have shown that wastewater treatment facilities are quite effective in microplastics removal from treated wastewater with total efficiencies between 90–98%; recent studies have reported removal rates of 95% (Talvitie *et al.*, 2017), 97% (Mintenig *et al.*, 2017) and 98% (Murphy *et al.*, 2016). The load of microplastics in treated wastewater which finally enters the environment is therefore not high but significant (Murphy *et al.*, 2016; Ziajahromi *et al.*, 2016).

The application of sewage sludge as fertilizer is a widespread practice in agriculture. Such agricultural use is made under a set of criteria to ensure that it does not adversely affect the quality of soils and the agricultural products, as well as the vegetation, livestock and people's lives (Kouloubis *et al.*, 2005). In Europe and North America, about 50% of total sewage sludge is used on agricultural fields (EPA, 2015; Eurostat, 2018). During wastewater treatment, more than 90% of microplastics are removed from water and most of this percentage remains in sludge (Bläsing & Amelung, 2018). It is important to mention that the largest disposal of sludge quantities is made onto agricultural land, applied as agricultural fertilizer and as soil improvers. It is estimated that, in Europe, the quantity of sewage sludge applied is between 4 and 5 million t, dry weight (Cieślik *et al.*, 2015; Willén *et al.*, 2016).

In discussing sources of microplastics in agricultural soils, it must be noted that several microplastics are added to agricultural soils other than by the application of WWTP by-products. For example, plastic mulches and polytunnels made of polyethylene (PE) are used to control the soil's temperature and moisture to create a microclimate for cultivation and to slow weed growth (Horton *et al.*, 2017). Subsequently, when these plastics are exposed to UV irradiation from sunlight, they are destroyed, lose their strength and become small fragments and microplastic particles (Sivan, 2011).

11.2.2 Microplastics quantity in agricultural soils from WWTP by-products

As previously mentioned, irrigation with wastewater accumulates harmful substances such as microplastics in agricultural soils. To evaluate the implications of the microplastics, the load which is deposited on soils during crop irrigation should be initially estimated.

Other than on the characteristics of each plant species, irrigation demand depends on climatic parameters such as temperature, precipitation and duration of the day, but also on soil parameters such as soil type and organic matter content.

The amount of microplastics that may reach wastewater-irrigated fields per cropping season per ha can be estimated by calculating a mean concentration value of microplastics in treated wastewater and then considering the mean water demand for cultivation of the plants grown.

According to EPA (2014), the influent water at WWTPs in West Ireland contained 97,000 particles m^{-3} whilst effluence samples (taken in May 2015) contained 2,000 particles m^{-3} (1,000 particles m^{-3} during tertiary treatment). These measurements show 2% of microplastics not being retained in the sludge and can be considered as a significant input of microplastic pollution. In other studies conducted in different countries, the concentration of microplastics in treatment wastewater was measured in Finland at 700–3,500 particles m^{-3} (Talvitie *et al.*, 2017), and in the Helsinki Region at 4,900 fibers and 8,600 microplastics m^{-3}; in France, it was measured at 14,000–15,000 particles m^{-3} (Dris *et al.*, 2015), in Germany at 0–9,400 particles m^{-3} (Mintenig *et al.*, 2017), in Holland at 20,000 particles m^{-3} (Leslie *et al.*, 2012) and 52,000 particles m^{-3} (Leslie *et al.*, 2013), and in Glasgow (Scotland) at 250 particles m^{-3} (Murphy *et al.*, 2016). These measurements show that despite efficient removal rates of microplastics by WWTPs, when applied to such a large volume of effluent in agriculture, significant amounts of microplastics are entering the environment (Mourgkogiannis *et al.*, 2018). From all the above-mentioned values, a concentration range of microplastics in treated wastewater of 0–52,000 particles m^{-3} was measured and 10,000 particles m^{-3} could be taken as the mean value. That value will be applied to estimate the microplastics load in soil after irrigation by treated wastewater. Table 11.1 shows the microplastics load which is added to soil as part of a general estimate of water needs for six crops under irrigation with wastewater.

As noted above, the highest concentration of WWTP microplastics has been found in sewage sludge. None of the sludge treatment technologies can remove these particles except sludge incineration which destroys them (Karapanagioti, 2017). The application of sewage sludge to agricultural land is governed in Europe by the European Union (EU) Sewage Sludge Directive (86/278/EEC) (EU, 1986) and in the US by USEPA 40 CFR 503 (USEPA, 1993) on the

Table 11.1 Estimation of microplastics (MP) load in soil after irrigation with treated wastewater.

Crop	Annual Irrigation Rates m^3 ha^{-1}	Mean MP Particles m^{-3} (1,000 s)	Annual MP Load in Soil ha^{-1} (million)
Maize	6,000	10	60
Cotton	5,500	10	55
Fodder–trefoil	9,000	10	90
Tomato	4,500	10	45
Cabbage	2,200	10	22
Olive trees	4,000	10	40

Table 11.2 Permitted quantities of dry sludge applied to soils in different countries (Kouloubis *et al.*, 2005).

Country	Average Annual Sludge Applied (t Dry Sludge ha^{-1} $year^{-1}$)	Maximum Sludge Applied (t Dry Sludge ha^{-1} $year^{-1}$)
Austria	2.5	5
Belgium	1–4	3–12
Denmark	10	100
France	3	30
Italy	2.5–5	7.5–15
Netherlands	1–10	1–10
Norway	2	20
Sweden	1	5
USA	10	

protection of the environment and especially for soil when sewage sludge is used in agriculture. However, neither the European nor the US regulations consider plastic as a potentially unwanted ingredient.

The detected plastic concentrations in sludge range from 1,500 to 24,000 particles kg^{-1}: 1,500–4,000 particles kg^{-1} (Zubris & Richards, 2005); 16,700 particles kg^{-1} (Magnusson & Noren, 2014); 100–24,000 particles kg^{-1} (Mintenig *et al.*, 2017) and 4,200–15,800 particles kg^{-1} (Mahon *et al.*, 2017). The quantities of sewage sludge applied by a number of different countries to agricultural soils ranges between 1–100 t dry sludge ha^{-1} (Table 11.2).

11.3 DEGRADATION OF MICROPLASTICS ON LAND

Generally, organic pollutants in soil are subject to complex degradation and transformation processes, the extents of which depend on many interdependent soil and climatic parameters. With the combined effect of microorganisms and soil fauna as well as various abiotic factors (pH, organic matter content, electrical conductivity etc) that affect them, a series of reactions take place and, finally, the various organic compounds are converted, under certain conditions and over long-term reactions, into volatile, water-soluble and solid products.

When exposed to environmental conditions including physical, biological and chemical processes, microplastics fragment over time to the size of nanoplastics. Generally, chemical substances are added to plastics during their manufacture to improve their properties, to optimize the applications and usage of products and to increase their shelf life (Roy *et al.*, 2011; Teuten *et al.*, 2009). These additives include plasticizers, antioxidants, flame retardants, ultraviolet stabilizers, lubricants and colorants, and they have an important environmental role as, among others, they increase the plastics degradation time and are potential pollutants due to their leaching and entry into the food chain. The removal of these additives from plastics primarily depends on the type of plastic to which they have been added, the size, the properties of the additive, and the environmental conditions which promote their degradation (Moore, 2008; Teuten *et al.*, 2009). The exposure of microplastics to conditions which promote biodegradation may create a relatively rapid fragmentation and, consequently, particles remain in the soil and are ultimately transferred to deeper soil layers.

In general, degradation of plastics refers to a chemical change in the molecular structure of the polymer which alters its properties. The efficiency of the different types of degradation is dependent on the chemical structure of the polymers. Polymers such as polyethylene (PE), polypropylene (PP), polystyrene (PS) and polyvinyl chloride (PVC) are resistant to hydrolytic and enzymatic degradation, with the consequential environmental accumulation of these materials.

It should be noted that, when microplastic particles are smaller, the ratio of surface to volume is higher and hence reactivity is higher. Consequently, the microplastics behaviour is more dynamic, acting as a suitable substrate for sorbing pollutants such as hydrated metals and organic compounds including PCBs, PAHs and organochlorinated compounds (pesticides) (Ng *et al.*, 2018).

Oxidative degradation is caused by free radicals generated when the materials are exposed, depending to a large extent on environmental conditions (e.g. ultraviolet radiation exposure, temperature, soil composition, humidity, oxygen) as well as to the chemical structure and crystallinity of the plastic (Fotopoulou & Karapanagioti, 2017; Nguyen, 2008). The degradation processes occur when the plastic is in contact with the ground.

Biodegradation is the biochemical process of mineralisation of an organic material by microorganisms and, finally, the production of CO_2 and H_2O (under aerobic conditions) or CO_2 and CH_4 (under anaerobic conditions) (Mohan, 2011). Biodegradation is affected by the properties of the plastic such as molecular weight, chemical structure, morphology, hydrophobicity and water absorption, and has an important role in the final form of plastics in the soil.

The interaction between microplastics and soil components is a dynamic process that involves a series of changes in natural biological and chemical properties. Microplastics are a composite mixture of polymers, catalysts and additives (Teuten *et al.*, 2009) which all influence their characteristics, behaviour and interactions with soil and organic minerals as well as with any agrochemicals (e.g. fertilizers, pesticides) present in the soil.

11.3.1 Additives in plastics

For a final plastic product, polymers are mixed with different additives to enhance their performance. During degradation of microplastics, these additives act as additional soil pollutants. The most common additives used in plastics production processes are phthalates, bisphenol A (BPA), flame retardants (FRs), polybrominated diphenyl ethers (PBDEs) and nonylphenols (NP).

11.3.1.1 Phthalates

Phthalates or phthalate esters are esters of phthalic acid. They are chemical compounds which are mainly used as plasticizers (substances added to plastics to increase their flexibility, durability and longevity). The most commonly used phthalates are di(2-ethylhexyl) phthalate (DEHP), dibutyl phthalate (DBP) and diethyl phthalate (DEP), and they are mainly used for PVC production (Net *et al.*, 2015). These phthalates cause concern because of indications that they act as endocrine disruptors, changing hormone levels. In 2008, the National Academy of Sciences (NAS) recommended that the effects of phthalates and other antiandrogens be investigated (Varshavsky *et al.*, 2016). According to EPA (2012), several human studies have reported associations of exposure to some phthalates with observed adverse reproductive outcomes including shortened anogenital distance observed in newborn boys, shortened pregnancies, lower sex and thyroid hormones, and reduced sperm quality observed in adults. Phthalates are easily released into the environment since they are not chemically bound to plastics, and they leach into the environment during the manufacture, use and disposal of plastics (Net *et al.*, 2015; Talsness *et al.*, 2009).

11.3.1.2 Bisphenol A

Bisphenol A (BPA) is an organic compound which is mainly used as an additive to other chemicals for making polycarbonate plastics. Plastics containing BPA exhibit excellent heat resistance, are rigid, light and transparent. Leaching of BPA into the environment can occur from food packaging debris (Sajiki & Yonekubo, 2003) or via untreated wastewater (Guerra *et al.*, 2015). The presence of BPA in the soil is an important problem and affects the symbiotic bacteria *Sinorhizobium meliloti* and therefore the amount of nitrogen at the roots of leguminous plants (Fox *et al.*, 2007); the reaction of soybean plants to the presence of BPA has been studied by several researchers (Sun *et al.*, 2013; Zhang *et al.*, 2016). Its effects on other crops, such as tomato (*Lycopersicum esculentum*), lettuce (*Lactuca sativa*), maize (*Zea mays*) and rice (*Oryza sativa*) have also been reported (Zhang *et al.*, 2016). These studies have shown that certain doses of BPA exposure could promote or inhibit growth, germination, pollen tube elongation, photosynthesis and hormone content in plants. Through the food chain in ecological systems, the hazards of BPA can extend to animals and even to humans (Jondeau-Cabaton *et al.*, 2013).

11.3.1.3 Flame retardants

Flame retardants (FRs) are used as a safety feature in plastic electronic devices, fabrics and many other plastic items to reduce their flammability. FRs include a wide range of chemicals, with the most commonly used compounds in plastic manufacture being organohalogen compounds such as polybrominated diphenyl ethers (PBDEs) and hexabromocyclododecane (HBCD). In 2004 and 2008, the EU banned several types of PBDEs (Betts, 2008) because of their association with endocrine disrupting effects, teratogenicity, and liver and kidney toxicity (Yogui & Sericano, 2009). Many flame retardants degrade into compounds which are toxic. Halogenated compounds with aromatic rings, for example, can degrade into dioxins and chlorinated dioxins which are also among highly toxic compounds. Bisphenol-A diphenyl phosphate (BADP) and tetrabromobisphenol A (TBBPA) can degrade to Bisphenol A (McCormick *et al.*, 2010). Organophosphorus compounds, another type of FR, have been detected in wastewater in Spain and Sweden (Marklund *et al.*, 2005; Rodil *et al.*, 2012) and in the Elbe River in Germany (Wolschke *et al.*, 2015).

11.3.1.4 Polybrominated diphenyl ethers

Polybrominated diphenyl ethers (PBDEs) are organobromine compounds which are used as flame retardant chemicals in a variety of commercial and household products such as electronic devices, electrical equipment, furniture, plastics, polyurethane foams, textiles and mattresses (EPA, 2017). The entrance of PBDEs into the environment may be through emissions from manufacturing processes, volatilization from various products which contain PBDEs, recycling wastes and landfill leachate (ATSDR, 2015). They are difficult to dissolve in water and bind strongly to soil particles or sediment. PBDEs remain in the environment for years without any significant degradation. Photolysis and pyrolysis can be the main factors of PBDEs transformation (Hutzinger & Thoma, 1987; Watanabe *et al.*, 1987). They have also been found in the air, soil, sediments, humans, wildlife, fish and other marine life, as well as in sewage treatment plant biosolids (Siddiqi *et al.*, 2003).

People are exposed to PBDEs through the food chain but they bioaccumulate in blood, breast milk and fat tissues (EPA, 2009). Household items contain PBDEs and high levels of PBDEs are present in indoor dust and in sewage sludge and effluents from wastewater treatment plants. PBDEs are endocrine disruptors and neurotoxins, and may cause chronic diseases, from cognitive disorder to hormonal and liver dysfunction (Siddiqi *et al.*, 2003).

11.3.1.5 Nonylphenols

Nonylphenols (NP) can cause estrogenic action and as endocrine disruptors are capable of interfering with the reproduction of numerous organisms. NPs are precursors to the non-ionic surfactants alkylphenol ethoxylates and nonylphenol ethoxylates, which are used in detergents, paints, pesticides, personal care products and plastics. Due to their physico–chemical characteristics and mainly because of their low solubility and high hydrophobicity, nonylphenols accumulate in environmental compartments which are characterized by high organic content, such as sewage sludge and river sediments (Soares *et al.*, 2008), and their main source is in treated wastewater effluent (Shinichi *et al.*, 2016). Months or longer may be required for its biodegradation in surface waters, soils and sediments. The degradation of NPs in soil depends on oxygen availability and other components in the soil; NPs mobility in soil is low (Soares *et al.*, 2008).

11.4 MICROPLASTICS AS AN AGRICULTURAL SOILS HAZARD

Microplastics can enter and potentially impact soil ecosystems, crops and livestock either as individual micro-items or through the toxic substances which are added during plastics manufacturing. Since there is the potential to affect all ecosystems (marine and terrestrial) as well as human health, it is important to investigate the pathways through which microplastics could be entering agro-ecosystems.

As a pollutant in soil, microplastics have several potential routes and factors which can determine their destination, including the microplastics properties such as size (Rillig *et al.*, 2017), hydrophobicity (Wan & Wilson, 1994), charge, density and shape (fiber, bead, foam), the soil's physico-chemical properties (Pachapur *et al.*, 2016), the soil's macropores, soil biota, agricultural activities, meteorological conditions and biological interactions. Soil aggregates can be embedded with microplastics, with the aggregates being formed and disintegrating, depending on soil characteristics, such as pH. Consequently, during formation of soil aggregates, microplastic particles, organic matter, and primary soil particles could be aggregated all together.

With agricultural activities such as plowing, microplastic particles can be moved into deeper soil layers. Also, harvesting of plants – especially plants which grow below the soil surface (e.g., beets and potatoes) – can also help to incorporate the microplastics into deeper soil layers with the shuffling of the soil. Macropores which are formed enhance the movement of particles as well as the movement of water which indirectly helps the particles move deeper in the soil; consequently, plant processes (e.g., root growth, uprooting) and soil-living fauna (e.g., earthworms, insects) can contribute to particle movement.

The wide range of plasticiser chemicals which plastics contain have already been discussed. Many of them have been identified as toxic or as endocrine disruptors. The places where microplastics accumulate in soil are possible sites where these chemicals are subsequently transferred to water, soil and soil-dwelling organisms. Studies have identified the presence of plasticiser chemicals and especially of the phthalate esters in agricultural soils: Zeng *et al.* (2008), Kong *et al.* (2012) and Wang *et al.* (2013) analyzed soil samples from farmland and identified phthalate compounds. The results suggest that plastic materials release chemicals to soil which may be taken up by plants (Sun *et al.*, 2015) and consequently enter the food chain and endanger human health. When microplastic particles move further into the soil profile, they eventually end up in groundwater, something which finally results in polluting subterranean waters and, consequently, the entire food chain with direct implications to human health.

Microplastics, have negative effects on organisms, mainly due to their accumulation in the gut or stomach, which affect organisms' behaviour and development (von Moos *et al.*, 2012; Watts *et al.*, 2016). Generally, studies have focused on the effects of microplastics on aquatic organisms in marine ecosystems; few studies have looked at effects on soil organisms, bioavailability, bioaccumulation or at a terrestrial ecological risk assessment.

Studies on the influence of microplastic ingestion by earthworms have shown that the microplastics may be fragmented internally by the organism into smaller particles and eventually return to the environment through defecation (Huerta Lwanga *et al.*, 2016). In other cases, at high exposure concentration, it has been observed that earthworms suffer, demonstrate weight loss and eventually die (Cao *et al.*, 2017). The only terrestrial species which have been exposed to microplastic particles under laboratory conditions, observing them for consequences on their life, are the earthworms Lumbricus terrestris (Huerta Lwanga *et al.*, 2016) and Eisenia andrei (Rodriguez-Seijo *et al.*, 2017), and nematodes, such as Daphnia magma, Thamnocephalus platyurus and Caenorhabditis elegans. All of these are sensitive to nanoplastics (Ng *et al.*, 2018). Nanoplastics are potentially more hazardous than microplastics due to fact that they can permeate biological membranes (Bouwmeester *et al.*, 2015; EFSA Panel on Contaminants in the Food Chain, 2016; Nel *et al.*, 2009). Earthworms, as part of soil fauna, influence a number of soil parameters, such as fertility and soil porosity. Consequently, earthworm–microplastic interactions affect soil quality and fertility.

The concentrations of microplastics on soil surfaces are currently unknown. Huerta Lwanga *et al.* (2016), for example, examined mortality in Lumbricus terrestris earthworms which were exposed to polyethylene particles; mortality was raised by 8% at a concentration of 450 g kg^{-1} polyethylene to 25% mortality at 600 g kg^{-1}. It is only possible to obtain estimates of the pollutant load evolution, since it is difficult to find real soils with such high microplastic concentration levels. However, it is important to evaluate the potential ecological implications of microplastic pollution at high concentrations as these concentrations are likely to increase with the fragmentation of plastics which already exist in the environment.

So far, there is no experimental evidence of micro- and nanoplastics being transferred from invertebrates to vertebrates; nevertheless, there is evidence of the transfer of microplastics from polluted land to vertebrates. According to Huerta Lwanga *et al.* (2017), chickens became polluted with plastic particles by their diet of earthworms.

Microplastics uptake by plants is not expected, due to the high molecular weight or large size of the plastic particles which prevents their penetration through plant cell walls. However, nanoplastics have been shown to enter plant cells: Bandmann *et al.* (2012) studied tobacco plants in cell culture (uptake dimensions of 20 and 40 nm nanopolystyrene); however, there are no studies on the translocation, storage and toxicity of nanoplastics in plants.

The capability to uptake, translocate and accumulate pollutants depends on plant species. Properties which affect the uptake of organic compounds are: the root properties (volume, density, surface area), xylem properties, transpiration, growth rate, water and lipid fractions, plasma membrane potential, tonoplast potential, cytoplasm and vacuoles pH (Trapp, 2000). Plants can metabolize a quantity of pollutants, including polychlorinated and polycyclic hydrocarbons (Sandermann, 1992). Generally, pollutants are stored as soluble and insoluble conjugates in plants. According to Calderón-Preciado *et al.* (2011), who studied plant uptake models and quantities of micropollutants in irrigation water, human exposure to 27 emerging micropollutants (including pharmaceuticals, fragrances and additives in plastic production including flame retardants and plasticizers) by vegetable and fruit consumption. Also, according to Torre-Roche *et al.* (2013), the different types of nanoparticles which are in soil and in applied wastewater could interact with pesticides in the soil, resulting in the increase or decrease of uptake of pesticides by different crops.

Another potential route to food chain contamination by microplastics is by their transfer into leaf crops through contact. However, this exposure parameter is considered negligible compared to other sources since, with basic hygiene (by washing vegetables), this risk can be avoided.

11.5 CONCLUSIONS

Globally, microplastics are appearing everywhere in aquatic and terrestrial ecosystems but the impacts of microplastic pollution in the environment are not well understood. Given that agricultural soils represent one of the largest environmental reservoirs of microplastics, it is important that they be studied and the ways in which they load the environment are understood, as well as the consequences of the existence of this pollution load. The factors, which determine the environmental fate of microplastics in agro-ecosystems are complex and more knowledge is required to completely understand their pathways and interactions, due to the multiple physical and biological forces that influence their transport mechanisms.

So far, based on existing studies and assessments, despite the application of wastewater and sewage sludge in agriculture, there appear to be no concentrations of microplastics with apparent impacts on both soil fauna and plants; concentrations for experimental assessments of pollution levels are quite large compared to the actual *in situ* values. Related to the effects on plant growth, there is no evidence for direct effects but only for indirect, such as the effect on soil quality, although this has been observed at high experimental concentrations.

The fact that no impact from microplastics pollution has so far been observed does not mean that the problem does not exist. The problem of the existence of microplastics in agro-ecosystems is real, just as its impact on marine ecosystems has become a serious problem. Consequently, it is now necessary, while agro-ecosystem concentrations of MPs are low and while there are currently no existing monitoring systems or risk assessment studies, to undertake research, make predictions for the future and propose solutions to the multifactorial MPs problem.

REFERENCES

Agency for Toxic Substances and Disease Registry (ATSDR) (2017). Toxicological Profile for Polybrominated Diphenyl Ethers. ATSDR, U.S. Department of Health and Human Services, Atlanta, GA.

Amec Foster Wheeler (2017). Intentionally Added Microplastics in Products–Final Report Prepared for the European Commission. Amec Foster Wheeler Environment & Infrastructure UK Ltd, London.

Bandmann V., Müller J. D., Köhler T. and Homann U. (2012). Uptake of fluorescent nano beads into BY2-cells involves clathrin-dependent and clathrin-independent endocytosis. *FEBS Letters*, **586**(20), 3626–3632.

Betts K. S. (2008). New thinking on flame retardants. *Environmental Health Perspectives*, **116**(5), 210–213.

Bläsing M. and Amelung W. (2018). Plastics in soil: analytical methods and possible sources. *Science of the Total Environment*, **612**, 422–435.

Bouwmeester H., Hollman P. C. and Peters R. J. (2015). Potential health impact of environmentally released micro- and nanoplastics in the human food production chain: experiences from nanotoxicology. *Environmetnal Science and Technology*, **49**, 8932–8947.

Browne M. A., Crump P., Niven S. J., Teuten E., Tonkin A., Galloway T. and Thompson R. (2011). Accumulation of microplastic on shorelines worldwide: sources and sinks. *Environmetnal Science and Technology*, **45**(21), 9175–9179.

Calderón-Preciado D., Matamoros V. and Bayona J. M. (2011). Occurrence and potential crop uptake of emerging contaminants and related compounds in an agricultural irrigation network. *Science of the Total Environment*, **412–413**, 14–19.

Cao D., Xiao W., Luo X., Liu G. and Zheng H. (2017). Effects of polystyrene microplastics on the fitness of earthworms in an agricultural soil. *IOP Conference Series: Earth and Environmental Science*, **61**, 12148.

Cieślik B. M., Namieśnik J. and Konieczka P. (2015). Review of sewage sludge management: standards, regulations and analytical methods. *Journal of Cleaner Production*, **90**, 1–15.

Dris R., Imhof H., Sanchez W., Gasperi J., Galgani F., Tassin B. and Laforsch C. (2015). Beyond the ocean: contamination of freshwater ecosystems with (micro-)plastic particles. *Environmental Chemistry*, **12**(5):539–550.

Duis K. and Coors A. (2016). Microplastics in the aquatic and terrestrial environment: sources (with a specific focus on personal care products), fate and effects. *Environmental Sciences Europe*, **28**(1), 2.

EFSA Panel on Contaminants in the Food Chain (2016). Presence of microplastics and nanoplastics in food, with particular focus on seafood. EFSA Journal, **14**(6), 04501. https://doi.org/10.2903/j.efsa.2016.4501.

Environmental Protection Agency (EPA) (2009). Polybrominated Diphenyl Ethers (PBDEs) Action Plan. EPA, United States.

Environmental Protection Agency (EPA) (2012). Phthalates, Action Plan. *Revised 03/14/2012*. EPA, United States.

Environmental Protection Agency (EPA) (2014). Scope, Fate, Risks and Impacts of Microplastic Pollution in Irish Freshwater Systems, Report No 210. EPA, United States.

Environmental Protection Agency (EPA) (2015). Urban Waste Water Treatment in 2014: A Report for the Year 2014. EPA, United States.

Environmental Protection Agency (EPA) (2017). Technical Fact Sheet: Polybrominated Diphenyl Ethers (PBDEs). EPA 505-F-17-015. EPA, United States.

European Union (EU) (1986). Council Directive on the Protection of the Environment, and in particular of the Soil, when Sewage Sludge is used in Agriculture (EU Directive 86/278/EEC). Council of the EU, Brussels.

Eurostat (2018). Sewage Sludge Production and Disposal. See: http://appsso.eurostat.ec.europa.eu/nui/show.do?lang=en&dataset=env_ww_spd (last accessed 09-04-2018).

Fotopoulou K. N. and Karapanagioti H. K. (2017). Degradation of Various plastics in the Environment. In: Hazardous Chemicals Associated with Plastics in the Marine Environment, (Handbook of Environmental Chemistry). H. Takada and H. K. Karapanagioti (ed.), DOI 10.1007/698_2017_11, Springer International Publishing.

Fox J. E., Gulledge J., Engelhaupt E., Burow M. E. and McLachlan J. A. (2007). Pesticides reduce symbiotic efficiency of nitrogen-fixing rhizobia and host plants. *Proceedings of the National Academy of Sciences*, **104**(24), 10282–10287.

GESAMP (2015). Chapter 3.1.2 Defining 'microplastics'. Sources, fate and effects of microplastics in the marine environment: a global assessment. In: (IMO/FAO/UNESCO-IOC/UNIDO/WMO/IAEA/UN/UNEP/UNDP Joint Group of Experts on the Scientific Aspects of Marine Environmental Protection (GESAMP)), P. J. Kershaw (ed.), Rep. Stud. GESAMP No. 90, 96p, London.

Guerra P., Kim M., Teslic S., Alaee M. and Smyth S. A. (2015). Bisphenol-A removal in various wastewater treatment processes: operational conditions, mass balance, and optimization. *Journal of Environmental Management*, **152**, 192–200.

Horton A. A., Walton A., Spurgeon D. J., Lahive E. and Svendsen C. (2017). Microplastics in freshwater and terrestrial environments: evaluating the current understanding to identify the knowledge gaps and future research priorities. *Science of the Total Environment*, **586**, 127–141.

Huerta Lwanga E., Gertsen H., Gooren H., Peters P., Salanki T., van der Ploeg M., Besseling E., Koelmans A. A. and Geissen V. (2016). Microplastics in the terrestrial ecosystem: implications for Lumbricus terrestris (Oligochaeta, Lumbricidae). *Environmental Science & Technology*, **50**(5), 2685–2691.

Huerta Lwanga E., Gertsen H., Gooren H., Peters P., Salánki T., van der Ploeg M., Besseling E., Koelmans A. A. and Geissen V. (2017). Incorporation of microplastics from litter into burrows of Lumbricus terrestris. *Environmental Pollution*, **220**, 523–531.

Hutzinger O. and Thoma H. (1987). Polybrominated dibenzodioxins and dibenzofurans. *Chemosphere*, **16**(89), 1877–1880.

Jondeau-Cabaton A., Soucasse A., Jamin E. L., Creusot N., Grimaldi M., Jouanin I., Aït-Aïssa S., Balaguer P., Debrauwer L. and Zalko D. (2013). Characterization of endocrine disruptors from a complex matrix using estrogen receptor affinity columns and high performance liquid chromatography–high resolution mass spectrometry. *Environmental Science and Pollution Research*, **20**(5), 2705–2720.

Kalavrouziotis I. K. and Koukoulakis P. H. (2011). Soil pollution under the effect of treated municipal wastewater. *Environmental Monitoring and Assessment*, **184**, 6297–6305.

Karapanagioti H. K. (2017). Microplastics and synthetic fibers in treated wastewater and sludge. In: Wastewater and Biosolids Management, I. K. Kalavrouziotis (ed.), IWA Publishing.

Kong S., Ji Y., Liu L., Chen L., Zhao X., Wang J., Bai Z. and Sun Z. (2012). Diversities of phthalate esters in suburban agricultural soils and wasteland soil appeared with urbanization in China. *Environmental Pollution*, **170**, 161–168.

Kouloubis P., Tsantilas C. and Gantidis N. (2005). Handbook of Good Agricultural Practice for the Valorisation of Sewage Sludge. Ministry of Agricultural Development and Food, Athens (in Greek).

Leslie H. A., Moester M., de Kreuk M. and Vethaak A. D. (2012). Verkennende studie naar lozing van microplastics door rwzi's. (Pilot study on emissions of microplastics from wastewater treatment plants) *H2O.*, **14/15**, 45–47.

Leslie H. A., van Velzen M. J. M. and Vethaak A. D. (2013). Microplastic Survey of the Dutch Environment. Novel Data Set of Microplastics in North Sea Sediments, Treated Wastewater Effluents and Marine Biota. Final report R-13/11. Institute for Environmental Studies, VU University, Amsterdam.

Magnusson K. and Noren F. (2014). Screening of Microplastic Particles in and Downstream a Wastewater Treatment Plant. IVL Swedish Environmental Research Institute.

Mahon A. M., O'Connell B., Healy M. G., O'Connor I., Officer R., Nash R. and Morrison L. (2017). Microplastics in sewage sludge: effects of treatment. *Environmental Science & Technology*, **51**(2), 810–818.

Marklund A., Andersson B. and Haglund P. (2005). Organophosphorus flame retardants and plasticizers in Swedish sewage treatment plants. *Environmental Science & Technology*, **39**(10), 7423–7429.

McCormick J., Paiva M. S., Häggblom M. M., Cooper K. R. and White L. A. (2010). Embryonic exposure to tetrabromobisphenol A and its metabolites, bisphenol A and tetrabromobisphenol A dimethyl ether disrupts normal zebrafish (Danio rerio) development and matrix metalloproteinase expression. *Aquatic Toxicology*, **100**(3), 255–62.

Mintenig S. M., Int-Veen I., Loder M. G. J., Primpke S. and Gerdts G. (2017). Identification of microplastic in effluents of waste water treatment plants using focal plane arraybased micro-Fourier-transform infrared imaging. *Water Research*, **108**, 365–372.

Mohan K. (2011). Microbial deterioration and degradation of polymeric materials. *Journal of Biochemical Technology*, **52**(2), 210–215.

Moore C. J. (2008). Synthetic polymers in the marine environment: a rapidly increasing, long-term threat. *Environmental Research*, **108**(2), 131–139.

Mourgkogiannis N., Kalavrouziotis I. K. and Karapanagioti H. K. (2018). Questionnaire-based survey to managers of 101 wastewater treatment plants in Greece confirms their potential as plastic marine litter sources. *Marine Pollution Bulletin*, **133**, 822–827.

Murphy F., Ewins C., Carbonnier F. and Quinn B. (2016). Wastewater treatment works (WwTW) as a source of microplastics in the aquatic environment. *Environmental Science & Technology*, **50**(11), 5800–5808.

Nel A. E., Mädler L., Velegol D., Xia T., Hoek E. M. V., Somasundaran P., Klaessig F., Castranova V. and Thompson M. (2009). Understanding biophysicochemical interactions at the nano-bio interface. *Nature Materials*, **8**, 543–557.

Net S., Sempéré R., Delmont A., Paluselli A. and Ouddane B. (2015). Occurrence, fate, behavior and ecotoxicological state of phthalates in different environmental matrices. *Environmental Science & Technology*, **49**(7), 4019–4035.

Ng E. L., Huerta Lwanga E., Eldridge S. M., Johnston P., Hu H. W., Geissen V. and Chen D. (2018). An overview of microplastic and nanoplastic pollution in agroecosystems. *Science of the Total Environment*, **627**, 1377–1388.

Nguyen T. Q. (2008). Polymer Degradation and Stabilization. Handbook of Polymer Reaction Engineering. Wiley-VCH, pp. 757–831.

Nizzetto L., Futter M. and Langaas S. (2016). Are agricultural soils dumps for microplastics of urban origin? *Environmental Science & Technology*, **50**(20), 10777–10779.

Pachapur V. L., Dalila Larios A., Cledón M., Brar S. K., Verma M. and Surampalli R. Y. (2016). Behavior and characterization of titanium dioxide and silver nanoparticles in soils. *Science of the Total Environment*, **563–564**, 933–943.

Papaioannou D., Kalavrouziotis I., Koukoulakis P., Papadopoulos F. and Psoma P. (2017). Metal fixation under soil pollution and wastewater reuse. *Desalination and Water Treatment*, **65**, 43–51.

Rillig M. C., Ziersch L. and Hempel S. (2017). Microplastic transport in soil by earthworms. *Scientific Reports*, **7**(1362).

Rios M. L. M., Karapanagioti H. and Ramirez A. N. (2018). Micro(nanoplastics) in the marine environment: current knowledge and gaps. *Current Opinion in Environmental Science & Health*, **1**, 47–51.

Rodil R., Quintana J., Concha-Graña E., López-Mahía P., Muniategui-Lorenzo S. and Prada-Rodríguez D. (2012). Emerging pollutants in sewage, surface and drinking water in Galicia (NW Spain). *Chemosphere.*, **86**(10), 1040–1049.

Rodriguez-Seijo A., Lourenço J., Rocha-Santos T. A. P., da Costa J., Duarte A. C., Vala H. and Pereira R. (2017). Histopathological and molecular effects of microplastics in Eisenia Andrei Bouché. *Environmental Pollution*, **220**(A), 495–503.

Roy P. K., Hakkarainen M., Varma I. K. and Albertsson A. C. (2011). Degradable polyethylene: fantasy or reality. *Environmental Science & Technology*, **45**(10), 4217–4227.

Sajiki J. and Yonekubo J. (2003). Leaching of bisphenol A (BPA) to seawater. *Chemosphere*, **51**(1), 55–62.

Sandermann H., Jr. (1992). Plant metabolism of xenobiotics. *Trends in Biochemical Sciences*, **17**(2), 82–84.

Shinichi M., Tomomi S. and Taisen I. (2016). Nonylphenol. In: Handbook of Hormones: Comparative Endocrinology for Basic and Clinical Research, T. Yoshio, A. Hironori and T. Kazuyoshi (eds.), Academic Press, pp. 573–574.

Siddiqi M. A., Laessig R. H. and Reed K. D. (2003). Polybrominated diphenyl ethers (PBDEs): new pollutants-old diseases. *Clinical Medicine & Research*, **1**(4), 281–90.

Sivan A. (2011). New perspectives in plastic biodegradation. *Current Opinion in Biotechnology*, **22**(3), 422–6.

Soares A., Guieysse B., Jefferson B., Cartmell E. and Lestera J. N. (2008). Nonylphenol in the environment: a critical review on occurrence, fate, toxicity and treatment in wastewaters. *Environment International*, **34**(7), 1033–1049.

Sun J., Wu X. and Gan J. (2015). Uptake and metabolism of phthalate esters by edible plants. *Environmental Science & Technology*, **49**(14), 8471–8478.

Sun H., Wang L. and Zhou Q. (2013). Effects of bisphenol A on growth and nitrogen nutrition of roots of soybean seedlings. *Environmental Toxicology and Chemistry*, **32**(1), 174–180.

Talsness C. E., Andrade A. J. M., Kuriyama S. N., Taylor J. A. and vom Saal F. S. (2009). Components of plastic: experimental studies in animals and relevance for human health. *Philosophical Transactions of the Royal Society, B*, **364**(1526), 2079–2096.

Talvitie J., Mikola A., Setala O., Heinonen M. and Koistinen A. (2017). How well is microlitter purified from wastewater? – a detailed study on the stepwise removal of microlitter in a tertiary level wastewater treatment plant. *Water Research*, **109**, 164–172.

Teuten E. L., Saquing J. M., Knappe D. R. U., Barlaz M. A., Jonsson S., Bjorn A., Rowland S. J., Thompson R. C., Galloway T. S., Yamashita R., Ochi D., Watanuki Y., Moore C., Viet P. H., Tana T. S., Prudente M., Boonyatumanond R., Zakaria M. P., Akkhavong K., Ogata Y., Hirai H., Iwasa I., Mizukawa K., Hagino Y., Imamura A., Saha M. and Takada H. (2009). Transport and release of chemicals from plastics to the environment and to wildlife. *Philosophical Transactions of the Royal Society. London B. Biological Sciences*, **364**(1526), 2027–2045.

Torre-Roche R. D. L., Hawthorne J., Deng Y., Xing B., Cai W., Newman L. A., Wang Q., Ma X., Hamdi H. and White J. C. (2013). Multiwalled carbon nanotubes and C60 fullerenes differentially impact the accumulation of weathered pesticides in four agricultural plants. *Environmental Science & Technology*, **47**(21), 12539–12547.

Trapp S. (2000). Modelling uptake into roots and subsequent translocation of neutral and ionisable organic compounds. *Pest Management Science*, **56**(9), 767–778.

USEPA (1993). Standards for the Use or Disposal of Sewage Sludge (USEPA 40 CFR 503). USEPA, Washington DC.

Varshavsky J. R., Zota A. R. and Woodruff T. J. (2016). A novel method for calculating potency-weighted cumulative phthalates exposure with implications for identifying racial/ethnic disparities among US reproductive-aged women in NHANES 2001–2012. *Environmental Science & Technology*, **50**(19), 10616–10624.

Von Moos N., Burkhardt-Holm P. and Köhler A. (2012). Uptake and effects of microplastics on cells and tissue of the blue mussel Mytilus edulis L. after an experimental exposure. *Environmental Science & Technology*, **46**(20), 11327–11335.

Wan J. and Wilson J. L. (1994). Colloid transport in unsaturated porous media. *Water Resources Research*, **30**(4), 857–864.

Wang J., Luo Y., Teng Y., Ma W., Christie P. and Li Z. (2013). Soil contamination by phthalate esters in Chinese intensive vegetable production systems with different modes of use of plastic film. *Environmental Pollution*, **180**, 265–273.

Watanabe I., Kashimoto T. and Tatsukawa R. (1987). Polybrominated diphenyl ethers in marine fish, shellfish and river sediments in Japan. *Chemosphere*, **16**(10–12), 2389–2396.

Watts A. J. R., Lewis C., Goodhead R. M., Beckett S. J., Moger J., Tyler C. R. and Galloway T. S. (2014). Uptake and retention of microplastics by the shore crab Carcinus maenas. *Environmental Science & Technology*, **48**(15), 8823–8830.

Watts A. J. R., Urbina M. A., Goodhead R., Moger J., Lewis C. and Galloway T. S. (2016). Effect of microplastic on the gills of the shore crab carcinus maenas. *Environmental Science & Technology*, **50**(10), 5364–5369.

Willén A., Junestedt C., Rodhe L., Pell M. and Jönsson H. (2016). Sewage sludge as fertiliser–environmental assessment of storage and land application options. *Water Science & Technology*, **75**(5–6), 1034–1050.

Wolschke H., Suhring R., Xie Z. and Ebinghaus R. (2015). Organophosphorus flame retardants and plasticizers in the aquatic environment: a case study of the Elbe River, Germany. *Environmental Pollution*, **206**, 488–493.

Yogui G. T. and Sericano J. L. (2009). Polybrominated diphenyl ether flame retardants in the U.S. marine environment: a review. *Environment International*, **35**(3), 655–666.

Zeng F., Cui K., Xie Z., Wu L., Liu M., Sun G., Lin Y., Luo D. and Zeng Z. (2008). Phthalate esters (PAEs): emerging organic contaminants in agricultural soils in peri-urban areas around Guangzhou, China. *Environmental Pollution*, **156**(2), 425–434.

Zhang D., Liu H., Qin X., Ma X., Yan C. and Wang H. (2016). The status and distribution characteristics of residual mulching lm in Xinjiang, China. *Journal of Integrative Agriculture*, **15**, 2639–2646.

Ziajahromi S., Neale P. A. and Leusch F. D. (2016). Wastewater treatment plant effluent as a source of microplastics: review of the fate, chemical interactions and potential risks to aquatic organisms. *Water Science & Technology*, **74**, 2253–2269.

Zubris K. A. V. and Richards B. K. (2005). Synthetic fibers as an indicator of land application of sludge. *Environmental Pollution*, **138**(2), 201–211.

Chapter 12

Possible effects of microplastics on human health

E. Sazakli and M. Leotsinidis

University of Patras, Lab of Public Health, Medical School, Patras, Greece

Keywords: Bisphenol A, Effects, Health impact, Nanoplastics, Particles, Phthalates, Toxicity

12.1 INTRODUCTION

Modern man, during an endless struggle to raise his standard of living and improve life, is trapped in a vicious circle in which the various products he creates are turning into a threat to his health at the end of their life cycle. A typical example of this is the expanded use of plastic products.

Plastic polymers are covalently bonded macromolecules of high relative molecular mass, composed after polymerization of many repeated subunits known as monomers. Polymers may be naturally occurring or synthetic. Plastic polymers are widely used to produce plastic products, fibers, coatings, adhesives and many other products (Lithner *et al.*, 2011). About 40% of the plastic manufactured is used for packaging, while other uses include building and construction (19.7%), automotive (10%) and other applications (16.7%) such as mechanical engineering, medical, furniture, etc. (Plastics Europe, 2017). In 2016, global plastic production reached the huge amount of 335 million tonne (t), with 60 t being processed in Europe alone (Plastics Europe, 2017). Of the seven plastic classification categories (Figure 12.1), which are commonly stamped on plastic items, only the first two are easily recyclable (Galloway, 2015).

Editors: Hrissi K. Karapanagioti and Ioannis K. Kalavrouziotis
doi: 10.2166/9781789061697_0177

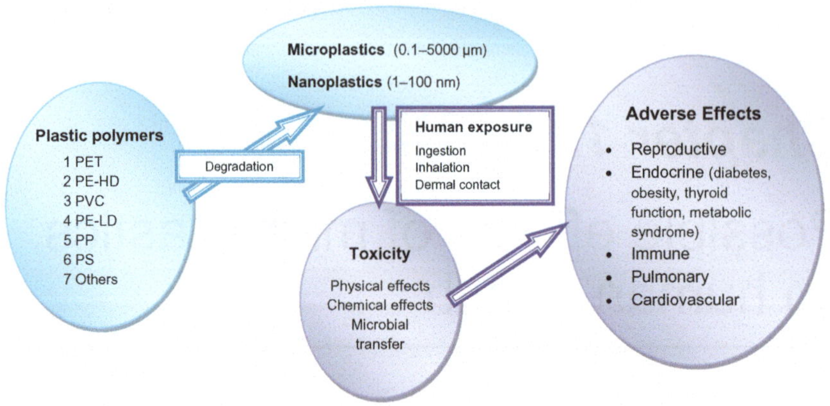

Figure 12.1 Flowchart showing the possible health effects of plastics on humans.

During production, initiators, catalysts, stabilizers, plasticizers and other additives are often needed in order to provide the desired properties of plastics, while at the same time residual monomers may be retained unreacted (Lithner *et al.*, 2011). Both additives and residual monomers are not bound to the polymer backbone and may easily migrate from the product to the surrounding environment, due to their low molecular weight. In addition, weathering processes and exposure to abiotic factors (such as UV light, heat, oxygen and mechanical abrasion) cause breaking of chemical bonds and depolymerisation. As a result, various compounds are distributed in the environment after disposal of plastic products.

12.2 HEALTH IMPACT

Potential hazardous effects to humans due to exposure to plastics derive from three areas: (a) toxicity due to exposure and accumulation of the particles, which leads to immune responses; (b) chemical effects caused by the intake of leached monomers, additives and other pollutants; and (c) microbial effects, owing to the ability of microplastic surfaces to transfer microbes (Wright & Kelly, 2017).

12.2.1 Particle effects

Microparticles (0.1–5,000 μm) and nanoparticles (1–100 nm) may derive from polymer-based materials either directly or after weathering and depolymerisation. Micro- and especially nanoparticles, exhibit substantially different physico-chemical properties from those presented by larger fragments of the same material. This differentiation creates the opportunity for increased uptake and interaction of nanoparticles with biological tissues – interactions not exhibited by the larger fragments of the same material (Nel *et al.*, 2006).

Human intake of micro- and nanoplastics takes place through oral, inhalation and dermal routes. Oral exposure can occur through ingestion of drinking water and marine products that have accumulated these particles, or directly through the actual ingestion of particles via other food items. Synthetic microfibers (>40 μm in length) have been found in honey and sugar, in average concentrations of 174 and 217 fibers $\cdot$ kg^{-1} honey and sugar, respectively (Liebezeit & Liebezeit, 2013). Polyethylene terephthalate (PET), polyethylene (PE) and cellophane, mostly small-sized (<200 μm), have been identified at concentrations of 550–681 particles $\cdot$ kg^{-1} in sea salt (Yang *et al.*, 2015). A dietary influx of nondegradable microparticles was estimated at 40 mg $\cdot$ person^{-1} $\cdot$ d^{-1}, which equates to a daily exposure of 10^{12-14} particles $\cdot$ person^{-1} (Powell *et al.*, 2010). Personal care products (toothpastes, scrubs, etc.) also constitute a source of oral exposure (Revel *et al.*, 2018).

After ingestion, particle uptake may occur via endocytosis by the M cells in Peyer's patches, or via paracellular persorption. Human studies have shown that microparticles of various types and sizes (from 0.1 to 150 μm) may further be translocated across the mammalian gut and enter the lymphatic system (Hussain *et al.*, 2001). Factors governing uptake and subsequent translocation to blood and lymph circulations are particle size, surface charge, hydrophobicity and presence of specific surface groups that could function as reactive sites (Galloway, 2015; Rist *et al.*, 2018; Wright & Kelly, 2017). Enhanced circulation times have been reported for hydrophilic and positively charged nanoparticles (Silvestre *et al.*, 2011). The liver and the spleen are the main secondary target organs but the kidneys and heart can also receive the uptaken particles (Galloway, 2015; Wright & Kelly, 2017). Unfortunately, nanoparticles are even capable of overcoming the blood-brain barrier, providing chemicals with direct access into the brain (Lockman *et al.*, 2004). Moreover, polyvinylchloride (PVC) particles have been observed to pass via the placenta into fetal circulation (Wright & Kelly, 2017). Microplastics are eliminated via bile (and finally excreted by feces), via urine, the pulmonary alveoli, peritoneal cavity, cerebrospinal fluid, and via milk in lactating women.

The wave action in aquatic environments, the application of sludge from wastewater treatment plants on land and atmospheric fallout produce airborne micro- and nanoplastics, which, in turn, may be inhaled by humans. Uptake and clearance of these particles depends on their size, shape and properties, the site of deposition, and the possible interactions between particles and biological structures. Normally, particles >1 μm will be subjected to mucociliary clearance via phagocytosis, whereas smaller particles may cross the epithelium and be deposited deeper in the lung (Wright & Kelly, 2017). Particle-induced Reactive Oxygen Species (ROS) generation has been documented and reported as an important mechanistic paradigm to explain the toxic effects of inhaled nanoparticles. This oxidative stress results in airway inflammation and interstitial fibrosis (Nel *et al.*, 2006).

Immunological response with associated cytokines release depends on the chemical structure of the plastic. PE particles (0.5–50 μm) provoke a non-immunological foreign body response, while PET particles (ranging from 0.5–20 μm) are stored in the cytoplasm and, if larger, locate extracellularly, causing substantial changes to the surrounding tissue (Wright & Kelly, 2017). Inflammation, genotoxicity, cell apoptosis and necrosis are some of the biological responses due to the cytotoxicity of particles which, if it persists, may cause tissue damage, fibrosis and carcinogenesis (Wright & Kelly, 2017). In addition, desorption of unbound chemicals, residual monomers and microbes that have been transferred from the particles' surface to the site of deposition may result in toxic effects. Such chemicals include PCBs, PAHs, metals, etc, many of which have known carcinogenetic, mutagenetic and teratogenetic properties.

Dermal exposure requires penetration of the stratum corneum, which is limited to particles <100 nm, so only nanoplastics are expected to be uptaken across skin (Revel *et al.*, 2018).

12.2.2 Chemical effects

The chemical action of the residual monomers and/or the additives of the micro- and nanoplastics, may induce adverse health effects such as reproductive toxicity (phthalates, Bisphenol A (BPA)), carcinogenicity (vinyl chloride, butadiene) and mutagenicity (benzene, phenol). Polyurethanes, PVC, epoxy resins and styrenic polymers are made of hazardous monomers classified as carcinogenic, mutagenic, or both (Wright & Kelly, 2017).

12.2.2.1 The case of Bisphenol A

One of the most studied compounds, with more than 10,000 studies dedicated to it, is BPA (4,4′-dihydroxy-2,2-diphenyl propane), the building block of polycarbonate plastic and epoxy resins. Despite the early recognition of its estrogenic activity dating back to 1936, the use of BPA in the plastics industy started in the 1950s and has continued to expand since then (Eladak *et al.*, 2015). The global volume consumption of BPA has been estimated at 7.7 million t in 2015 and is projected to reach 10.6 million t by 2022.

As is often the case with toxic chemicals, BPA has been the focus of a decade-long scientific controversy. This debate has reshaped environmental health studies, in the way that it raised the issue of extrapolation of experimental results to low-dose effects, the importance of considering critical periods of exposure in experimental design, and the existence of non-monotonic dose response curves (Eladak *et al.*, 2015; Vandenberg *et al.*, 2009). Such curves are either "U-shaped" or "inverted-U–shaped", meaning that low doses can lead to effects unpredictable to (and sometimes opposite to) those from high-dose experiments (Myers *et al.*, 2009).

Due to its two benzene rings and two (4, 4)-OH substituents, BPA fits in to the estrogen receptor binding pocket. Kinetic studies have determined that BPA binds to both estrogen receptors ERα and ERβ with approximately 10-fold higher affinity to ERβ (Halden, 2010; Vandenberg *et al.*, 2009). Even though the binding affinity of BPA is ~10,000-fold lower than the compound that it imitates (estradiol), the discovery that BPA at nM doses promotes calcium influx in MFC-7 breast cancer cells in the same way as estradiol classifies BPA as an estrogen and endocrine disrupting chemical (EDC) which interferes with normal hormone biosynthesis, signaling or metabolism (vom Saal & Hughes, 2005).

The latest advances in the field of male reproductive function quality indicate a global decrease in sperm count that has occurred over the past five decades, along with a steady increase in the incidence of testicular cancer in young men. Both these disorders may occur due to faults in testis development during fetal life. In order for masculinisation to evolve normally, androgens must act and it is here that the EDCs have a negative effect. Associations between masculinization defects and BPA exposure during fetal life have been observed in China, where sons of workers who were occupationally exposed to BPA during pregnancy showed a decreased anogenital distance (the distance from the anus to the genitalia) (Miao *et al.*, 2011), in Korea, where BPA plasma levels were significantly higher in newborn boys with hypospadias than in newborns without hypospadias (Choi *et al.*, 2012), and in France, where BPA was related to undescended testes in newborn boys (Fenichel *et al.*, 2013). A significant finding is the species-specific action of BPA. While concentrations of BPA as low as $10\,\text{nmol}\cdot\text{L}^{-1}$ ($2.28\,\text{ng}\cdot\text{mL}^{-1}$) reduce the fetal Leydig cell-specific functions in human fetal testes, at least 100-fold higher concentrations are required in mouse and rat testes (N'Tumba-Byn *et al.*, 2012).

In adult women with Polycystic Ovaries (PCO), blood BPA levels are higher compared to controls and a statistically significant positive association exists between circulating androgen levels and BPA (Kandaraki *et al.*, 2011). However, human epidemiological studies do not confirm an association between BPA and endometrial disorders or breast cancer, even though animal studies have shown that prenatal exposure to BPA causes disruption of the mammary tissue and increases susceptibility of the tissue to chemical carcinogens (Rochester, 2013).

Although some studies demonstrate positive correlations between maternal BPA exposure and birth weight, others find negative or no associations (Rochester, 2013). Concerning childhood behaviour and neurodevelopment, a detailed review undertaken by Mustieles *et al.* (2015) suggested that BPA is strongly associated with neurobehavioural problems (hyperactivity, aggression, intelligence, memory) in children exposed to BPA *in utero* or before puberty, indicating disruption of the brain during "critical developmental windows". These effects seem to be sex-dependent and may be mediated via endocrine-related mechanisms, epigenetic modulations or synaptogenic alterations. Prenatal and postnatal BPA exposure seems to play a role in the development of asthma; however, this needs to be confirmed by additional longitudinal studies.

Obesity, metabolic syndrome and diabetes are some of the disorders linked to BPA exposure and the reason given for a total BPA prohibition recommended by endocrinologists and diabetologists. Adiponectin is an adipocyte-specific hormone that protects against metabolic syndrome. Suppression of adiponectin release and stimulation of the secretion of interleukin-6 and TNFα implicate BPA in causing insulin resistance, lipid accumulation and increasing susceptibility to the development of metabolic syndrome (Hugo *et al.*, 2008). Positive correlations, not necessarily causal, between BPA exposure and obesity have been observed in cross-sectional epidemiological studies. Similar correlations found between urinary BPA and either self-reported or diagnosed diabetes remain to be confirmed by prospective longitudinal studies (Fenichel *et al.*, 2013).

Cardiovascular disorders and hypertension are adult onset diseases that have been associated with adult BPA exposure, as found in studies mostly coming from the National Health and Nutrition Examination Survey (NHANES) data. While participants had reported several diseases, significant associations of elevated urinary BPA were only found with coronary heart disease and diabetes, as well as with altered liver function in adults (elevated alkaline phosphatase and lactate dehydrogenase) (Lang *et al.*, 2008; Melzer *et al.*, 2010).

Finally, thyroid function may also be disrupted (higher triidothyronine (T3) and lower thyroid stimulating hormone (TSH)) by BPA exposure in a complex way, as both agonistic and antagonistic interactions with the thyroid receptor have been shown in human studies (Rochester, 2013). Other health outcomes possibly related with BPA exposure are immune function, albuminuria, chronic inflammation, oxidative stress and altered epigenetic markers, and gene expression (Rochester, 2013).

Concerning human metabolism of BPA, Völkel *et al.* (2002) found that after oral exposure BPA was rapidly metabolized in the liver to form the inactivated BPA–glucuronide, immediately eliminated by renal excretion. Models investigating BPA kinetics indicate that rats and humans may act differently concerning BPA clearance rates, intestinal glucuronidation and excretion rates. Moreover, these models do not examine chronic low-dose exposures that occur in the environment (Vandenberg *et al.*, 2009). Biomonitoring studies have confirmed a broad human exposure to BPA which was detectable in the urine of almost all adults and children tested, in the serum of pregnant women, breast milk, follicular and amniotic fluid, cord blood and placental tissue, and human fetal livers. The urinary/serum BPA concentrations found in the general population ranged from 0.4 to 9 $\mu g \cdot L^{-1}$ (Rochester, 2013). It is noteworthy that the tissues that exhibited the highest BPA concentrations – up to the level of 11.2 $\mu g \cdot L^{-1}$ – are those related to embryo development and the maternal influence of postnatal development of infants (Pjanic, 2017).

European and US official agencies, after a series of evaluations of effects, reviews of the scientific literature, and meetings and discussions between experts, concluded that current BPA levels present no risk to the general population (Tyl, 2014). The Harvard Center for Risk Analysis, under a formal deliberation framework, reported no consistent affirmative evidence of potential developmental and reproductive toxicity of BPA in animals at doses well below the US Reference Dose (RfD: $0.05\ mg \cdot kg^{-1} \cdot d^{-1}$) (Gray *et al.*, 2004). By setting seven criteria and reviewing the data, the Harvard Panel concluded that two large multigenerational studies provided the most relevant and reliable data, as they used a large number of animals, a wide distribution of doses, a number of endpoints and followed good laboratory practices. The official agencies stated that positive findings from some explorative studies have not been confirmed in subsequent studies with higher numbers of animals and "rodent data can well be used as a basis for human risk evaluation" (Hengstler *et al.*, 2011). However, this and similar reports accepted a lot of criticism as they were based on studies funded by the American Plastics Council and the Society of the Plastics Industry (Vogel, 2009). In June 2008, the National Toxicology Program drew final conclusions about BPA risk assessment, reporting that "there is some concern for neural and behavioral effects and the prostate gland in fetuses, infants, and children at current human exposures" (Vandenberg *et al.*, 2009). In 2011, the Advisory Committee of the German Society of Toxicology assessed the previous Tolerable Daily Intake (TDI: $0.05\ mg \cdot kg^{-1} \cdot d^{-1}$), as adequately justified and declared that BPA exposure represents no noteworthy risk to the health of the human population, including newborns and babies (Hengstler *et al.*, 2011). However, in 2015, the European Food Safety Authority (EFSA) itself issued a temporarily reduced TDI of $4\ \mu g \cdot kg^{-1} \cdot d^{-1}$ ($0.004\ mg \cdot kg^{-1} \cdot d^{-1}$), based on new data and a refined risk assessment, considering the uncertainty in the database regarding mammary glands and reproductive, metabolic, neurobehavioural and immune systems. In September 2018, the EFSA started a re-evaluation of the temporary TDI which is anticipated to be ready by 2020.

12.2.2.2 The case of phthalates

Phthalates are diesters of phthalic acid and represent a group of compounds belonging to plasticizers, due to their ability to provide flexibility and elasticity to plastics. Phthalates are ubiquitous in the environment as they are used in a large number of products. It is estimated that 2 million $t \cdot y^{-1}$ of di-2-ethylhexyl phthalate (DEHP) is produced for a variety of industrial products and medical devices, while di-ethyl phthalate (DEP), di-butyl phthalate (DBP) and di-methyl phthalate (DMP) are primarily used in cosmetics, personal care products and as an enteric coating for medicinal tablets (Sathyanarayana, 2008). Intravenous and respiratory tubing, extracorporeal membrane oxygenation tubes, gloves and nasogastric tubes are some of the medical devices that can contain 20–40%

DEHP by weight. Because of the non-covalent bond of phthalates to the plastic matrix, these compounds can easily leach out of tubing, especially when heated (as with warm saline/blood). The TDI value of DEHP is set to $50\,\mu g \cdot kg^{-1} \cdot d^{-1}$ (Testai *et al.*, 2016). Exposure to DEHP may significantly exceed the TDI in some specific groups, among which are adult patients undergoing haemodialysis (who may present median exposure levels exceeding the TDI by 2–12 fold) and premature neonates in intensive care units (who may receive DEHP levels up to $6000\,\mu g \cdot kg^{-1} \cdot d^{-1}$) (Testai *et al.*, 2016). Food and food products, as well as indoor air and house dust are other sources for human exposure to phthalates. Measurable concentrations of phthalates were observed in 72% of personal care products such as hair gels, deodorant, perfumes and hair sprays. Phthalates are also detected in children's plastic toys (Sathyanarayana, 2008). It has been reported that children are more exposed to phthalates than adults because of their hand-to-mouth activity, their larger surface area to weight ratio and their enhanced metabolic rate (Mariana *et al.*, 2016). A German cohort documented that exposure to High-Molecular Weight (HMW) phthalates appears to be driven by dietary intake, while non-dietary routes (like personal care products, dust and indoor air) appear to explain exposure to Low-Molecular Weight (LMW) phthalates (Koch *et al.*, 2013). Due to the extensive scientific and public awareness, the use of DEHP in toys for children under 3 years of age is not allowed in the European Union (EU), while the US government forbade their use in amounts >0.1% in children's toys and childcare articles (Mariana *et al.*, 2016). *In utero* exposures to DEHP phthalates may determine future health effects (Sathyanarayana, 2008), as they may disturb normal balance between androgens and estrogens (Talsness *et al.*, 2009).

After exposure, phthalate diesters are rapidly metabolized to their respective monoesters, which are the major and more bioactive metabolites. Phthalates have short biological half-lives, from hours to days, and are quickly excreted from the body. Certain phthalate diesters and their metabolites are measurable in human breast milk, cord blood and other pregnancy-related specimens. However, the biomarkers of choice are the phthalates monoesters in urine used for estimating phthalate exposures in epidemiologic studies (Mariana *et al.*, 2016). Moreover, the metabolic pathway of DEHP in humans is qualitatively independent of the exposure route (Testai *et al.*, 2016).

Human studies investigating the effect of phthalates exposure on reproductive development are limited compared to those with animal models but they evidence antiandrogenic activity, disruption of normal endocrine function and possible estrogenic actions (Sathyanarayana, 2008). In contrast to other anti-androgens, phthalates mainly inhibit fetal testicular testosterone biosynthesis (Talsness *et al.*, 2009). Testicular dysgenesis syndrome (characterized by a number of reproductive disorders including cryptorchidism, hypospadias and smaller reproductive organs) has been associated with *in utero* exposure to EDCs and could lead to reduction in semen quality, infertility and an increased risk for testicular cancer (Halden, 2010). Increased prenatal concentrations of urinary phthalate metabolites in mothers have been correlated with decreased anogenital distance (used as a marker of androgenization), as well as with penile width and testicular descent in male infants (Swan, 2008). However, later studies (referenced in Mariana *et al.*, 2016) have reached inconsistent results regarding maternal phthalate exposure and decreased anogenital distance, hypospadias or cryptorchidism, so a definite conclusion cannot, so far, be drawn. Contradictory results have also been reported for the effect of phthalate exposure on semen parameters in male humans; however, most of the evidence supports a correlation between DEHP and DBP exposure with lower semen quality (Mariana *et al.*, 2016). Increased serum DEHP levels were found in young girls with premature thelarche (premature breast development) in a study which, however, suffers from several methodological issues (Colon *et al.*, 2000). More recent and larger studies (Frederiksen *et al.*, 2012; Wolff *et al.*, 2010) did not confirm an association between phthalate exposure and precocious puberty. In all, there are indications that phthalates speed up pubertal onset in girls and delay pubertal development in boys, while they may play a role in the manifestation of childhood obesity (Katsikantami *et al.*, 2016). Prenatal exposure to phthalates has been linked to reduced gestational time and low birth weight; however, epidemiological studies have yielded conflicting results and mechanisms are poorly understood (Mariana *et al.*, 2016).

Concerning cardiovascular diseases, a small body of evidence indicates that increased phthalates concentrations may correlate with an increased risk of coronary heart disease, atherosclerosis, increased diastolic blood pressure and pregnancy-induced hypertensive diseases; however, this field should be further investigated given that in animal models the adverse effects of DEHP on cardiomyocyte function has already been proven (Mariana *et al.*, 2016).

With regard to thyroid hormones, which are critical for regulation of growth and metabolism, evidence exists about phthalates disrupting normal thyroid function. In adults, an inverse association between mono-2-ethylhexyl phthalate (MEHP) urinary concentrations and free T4 and T3 serum levels has been observed (Meeker *et al.*, 2007), while in another study (of pregnant Taiwanese women) an increased second trimester phthalate exposure was associated with maternal hypothyroidism (Huang *et al.*, 2007).

Other developmental effects of phthalate exposures may include pulmonary system effects, such as allergies, rhinitis, asthmatic reactions and direct toxicity (Meeker *et al.*, 2009). DEHP, when inhaled and locally hydrolyzed to MEHP, has been implicated as causing an increased risk of airway inflammation due to its mimicking of prostaglandins and thromboxanes. In a recent study assessing the impact of phthalates exposure of mothers on the early neurodevelopmental performance of their offspring at 13–24 months of age, associations were observed between monoethyl phthalate (MEP) in maternal urine and DEHP metabolite in breast milk with early adverse mental development (Kim *et al.*,

2018). Many studies highlight the sex-specific adverse effects of phthalates exposure, suggesting that males and females are affected by different phthalates and in different ways (Katsikantami *et al.*, 2016; Kim *et al.*, 2018).

12.2.2.3 Other additives

A plethora of substances and additives are used in the synthesis of plastic products. These include antioxidants, UV stabilizers, surfactants, pigments, dispersants, lubricants, anti-statics, nanofibers, biocides and fragrances. PVC has heat stabiliser additives to keep the polymer stable during production and plasticizers such as phthalates to allow flexibility. UV stabilizers and antioxidants are added to polypropylene (PP) which, otherwise, is vulnerable to oxidation (Lithner *et al.*, 2011). Brominated flame retardants, such as polybrominated diphenyl ether (PBDE) and tetrabromobisphenol A (TBBPA), may leach from acrylonitrile-butadiene-styrene (ABS) which is used in electronics, appliances and fabrics; both compounds have been shown to disrupt thyroid hormone homeostasis, while PBDEs also exhibit anti-androgen action (Rist *et al.*, 2018; Talsness *et al.*, 2009). PBDEs were found in household dust in concentrations $>90\ ng \cdot g^{-1}$ dust (Wright & Kelly, 2017).

Other additives of concern to human health include nonylphenol added to polyolefins, the antibacterial and antifungal agent triclosan, the UV screen and printing ink additive benzophenone, and organotins used as heat stabilizers (Galloway, 2015; Lithner *et al.*, 2011).

Lithner *et al.* (2011) conducted a comprehensive hazard ranking of plastic polymers based on physical, environmental and health risks originating from their constituent monomers. The polymer types that received the highest hazard rankings were Polyurethane as a flexible foam, Polyacrylamide and PVC. Specifically, PVC may cause harm through the inhalation of the PVC granules in conjunction with the slow release of the carcinogenic vinylchloride from the particles to adjacent lung tissue; thus, both inflammation and carcinogenic risks are linked to PVC (Prata, 2018). Styrene oligomers, released from polystyrene (PS) plastic articles, are suspected of exhibiting estrogen-like activity and causing ROS production (Halden, 2010). Two other hazardous raw material substances are benzene and butadiene, which are both classified as carcinogenic and mutagenic (Lithner *et al.*, 2011).

12.2.3 Microbial transfer

Finally, microplastics can act as vectors for potential pathogens. The surface of microplastics is ideal for microbial colonization; well-developed biofilms are established and remain on the surface of different types of plastics, such as PE and PET. Given their resistance, microorganisms may be directly transported to human tissue (in the gastrointestinal tract or lung) and alter the physiological community of the tissue microbiome. In this way, defence mechanisms may be circumvented and, as a result, infection and other immune responses could occur, especially in debilitated areas already suffering from particle toxicity (Prata, 2018; Wright & Kelly, 2017). Potentially pathogenic *Vibrio parahaemolyticus* have been identified on microplastics made of PE, PP, and PS (Revel *et al.*, 2018).

12.3 CONCLUSIONS

Micro- and nanoplastics attack almost every single tissue, organ, organism and, eventually, the whole biosphere. For the impact of microplastics on human health, there are still many questions pending. At the same time, increased public awareness leads occasionally to exaggerated reactions not actually based on scientific findings. To avoid response bias, risk assessment models must be employed and results should be communicated to the general public by experts in the field (Kontrick, 2018). The expertise of medical toxicologists will contribute to efficient future actions. In addition, focus should be directed towards creating sustainable means of production, use and disposal of plastic materials.

REFERENCES

Choi H., Kim J., Im Y., Lee S. and Kim Y. (2012). The association between some endocrine disruptors and hypospadias in biological samples. *Journal of Environmental Science and Health A Hazardous Substances & Environmental Engineering*, **47**(13), 2173–2179.

Colon I., Caro D., Bourdony C. J. and Rosario O. (2000). Identification of phthalate esters in the serum of young Puerto Rican girls with premature breast development. *Environmental Health Perspectives*, **108**(9), 895–900.

Eladak S., Grisin T., Moison D., Guerquin M. J., N'Tumba-Byn T., Pozzi-Gaudin S., Benachi A., Livera G., Rouiller-Fabre V. and Habert R. (2015). A new chapter in the bisphenol A story: bisphenol S and bisphenol F are not safe alternatives to this compound. *Fertility and Sterility*, **103**(1), 11–21.

Fenichel P., Chevalier N. and Brucker-Davis F. (2013). Bisphenol A: an endocrine and metabolic disruptor. *Ann Endocrinol (Paris)*, **74**(3), 211–220.

Frederiksen H., Sorensen K., Mouritsen A., Aksglaede L., Hagen C. P., Petersen J. H., Skakkebaek N. E., Andersson A. M. and Juul A. (2012). High urinary phthalate concentration associated with delayed pubarche in girls. *International Journal of Andrology*, **35**(3), 216–226.

Galloway T. S. (2015). Micro- and nano-plastics and human health. In: Marine Anthropogenic Litter. M. Bergmann, L. Gutow and M. Klages (eds.), Springer, pp. 344–366.

Gray G. M., Cohen J. T., Cunha G., Hughes C., McConnell E. E., Rhomberg L., Sipes I. G. and Mattison D. (2004). Weight of the evidence evaluation of low-dose reproductive and developmental effects of bisphenol A. *Human and Ecological Risk Assessment*, **10**(5), 875–921.

Halden R. U. (2010). Plastics and health risks. *Annuual Review of Public Health*, **31**, 179–194.

Hengstler J. G., Foth H., Gebel T., Kramer P. J., Lilienblum W., Schweinfurth H., Völkel W., Wollin K. M. and Gundert-Remy U. (2011). Critical evaluation of key evidence on the human health hazards of exposure to bisphenol A. *Critical Reviews in Toxicology*, **41**(4), 263–291.

Huang P. C., Kuo P. L., Guo Y. L., Liao P. C. and Lee C. C. (2007). Associations between urinary phthalate monoesters and thyroid hormones in pregnant women. *Human Reproduction*, **22**(10), 2715–2722.

Hugo E. R., Brandebourg T. D., Woo J. G., Loftus J., Alexander J. W. and Ben-Jonathan N. (2008). Bisphenol A at environmentally relevant doses inhibits adiponectin release from human adipose tissue explants and adipocytes. *Environ Health Perspect*, **116**(12), 1642–1647.

Hussain N., Jaitley V. and Florence A. T. (2001). Recent advances in the understanding of uptake of microparticulates across the gastrointestinal lymphatics. *Advanced Drug Delivery Reviews*, **50**(1–2), 107–142.

Kandaraki E., Chatzigeorgiou A., Livadas S., Palioura E., Economou F., Koutsilieris M., Palimeri S., Panidis D. and Diamanti-Kandarakis E. (2011). Endocrine disruptors and polycystic ovary syndrome (PCOS): elevated serum levels of bisphenol A in women with PCOS. *The Journal of Clinical Endocrinology and Metabolism*, **96**(3), E480–E484.

Katsikantami I., Sifakis S., Tzatzarakis M. N., Vakonaki E., Kalantzi O. I., Tsatsakis A. M. and Rizos A. K. (2016). A global assessment of phthalates burden and related links to health effects. *Environment International*, **97**, 212–236.

Kim S., Eom S., Kim H-J., Lee J. J., Choi G., Choi S., Kim S. Y., Cho G., Kim Y. D., Suh E., Kim S. K., Kim G. H., Moon H. B., Park J., Choi K. and Eun S. H. (2018). Association between maternal exposure to major phthalates, heavy metals and persistent organic pollutants, and the neurodevelopmental performances of their children at 1 to 2 years of age-CHECK cohort study. *Science of the Total Environment*, **624**, 377–384.

Koch H. M., Lorber M., Christensen K. L., Pälmke C., Koslitz S. and Brüning T. (2013). Identifying sources of phthalate exposure with human biomonitoring: results of a 48 h fasting study with urine collection and personal activity patterns. *International Journal of Hygiene and Environmental Health*, **216**(6), 672–681.

Kontrick A. (2018). Microplastics and human health: our great future to think about now. *Journal of Medical Toxicology*, **14**(2), 117–119.

Lang I. A., Galloway T. S., Scarlett A., Henley W. E., Depledge M., Wallace R. B. and Melzer D. (2008). Association of urinary bisphenol A concentration with medical disorders and laboratory abnormalities in adults. *JAMA*, **300**(11), 1303–1310.

Liebezeit G. and Liebezeit E. (2013). Non-pollen particulates in honey and sugar. *Food Additives & Contaminants Part A Chemistry, Analysis, Control, Exposure & Risk Assessment*, **30**(12), 2136–2140.

Lithner D., Larsson A. and Dave G. (2011). Environmental and health hazard ranking and assessment of plastic polymers based on chemical composition. *Science of the Total Environment*, **409**(18), 3309–3324.

Lockman P. R., Koziara J. M., Mumper R. J. and Allen D. D. (2004). Nanoparticle surface charges alter blood-brain barrier integrity and permeability. *Journal of Drug Targeting*, **12**(9–10), 635–641.

Mariana M., Feiteiro J., Verde I. and Cairrao D. (2016). The effects of phthalates in the cardiovascular and reproductive systems: a review. *Environment International*, **94**, 758–776.

Meeker J. D., Sathyanarayana S. and Swan S. H. (2009). Phthalates and other additives in plastics: human exposure and associated health outcomes. *Philosophical Transactions of the Royal Society of London B Biological Sciences*, **364**, 2097–2113.

Meeker J. D., Calafat A. M. and Hauser R. (2007). Di (2-ethylhexyl) phthalate metabolites may alter thyroid hormone levels in men. *Environmental Health Perspectives*, **115**(7), 1029–1034.

Melzer D., Rice N. E., Lewis C., Henley W. E. and Galloway T. S. (2010). Association of urinary bisphenol A concentration with heart disease: evidence from NHANES 2003/06. *PLOS ONE*, **5**(1), e8673.

Miao M., Yuan W., He Y., Zhou Z., Wang J., Gao E., Li G. and Li D. K. (2011). In utero exposure to bisphenol-A and anogenital distance of male offspring. *Birth Defects Research A Clinical and Molecular Teratology*, **91**, 867–72.

Mustieles V., Pérez-Lobato R., Olea N. and Fernández M. F. (2015). Bisphenol A: human exposure and neurobehavior. *Neurotoxicology*, **49**, 174–184.

Myers J. P., Zoeller R. T. and vom Saal F. S. (2009). A clash of old and new scientific concepts in toxicity, with important implications for public health. *Environmental Health Perspectives*, **117**(11), 1652–1655.

N'Tumba-Byn T., Moison D., Lacroix M., Lecureuil C., Lesage L., Prud'homme S. M., Pozzi-Gaudin S., Frydman R., Benachi A., Livera G., Rouiller-Fabre V. and Habert R. (2012). Differential effects of bisphenol A and diethylstilbestrol on human, rat and mouse fetal Leydig cell function. *PLOS ONE*, **7**(12), e51579.

Nel A., Xia T., Mädler L. and Li N. (2006). Toxic Potential of Materials at the Nanolevel. *Science*, **311**(5761), 622–627.

Pjanic M. (2017). The role of polycarbonate monomer bisphenol-A in insulin resistance. *Peer J*, **5**, e3809.

Plastics Europe (2017). Plastics – The Facts 2017, an Analysis of European Plastics Production, Demand and Waste Data. Plastics Europe, Brussels, Belgium. Available at: https://www.plasticseurope.org/application/files/5715/1717/4180/Plastics_the_facts_2017_FINAL_for_website_one_page.pdf.

Powell J. J., Faria N., Thomas-McKay E. and Pele L. C. (2010). Origin and fate of dietary nanoparticles and microparticles in the gastrointestinal tract. *Journal of Autoimmunity*, **34**(3), J226–J233.

Prata J. C. (2018). Airborne microplastics: consequences to human health? *Environmental Pollution*. **234**, 115–126.

Revel M., Châtel A. and Mouneyrac C. (2018). Micro(nano)plastics: a threat to human health? *Current Opinion in Environmental Science & Health*, **1**, 17–23.

Rist S., Carney Almroth B., Hartmann N. B. and Karlsson T. M. (2018). A critical perspective on early communications concerning human health aspects of microplastics. *Science of the Total Environment*, **626**, 720–726.

Rochester J. R. (2013). Bisphenol A and human health: a review of the literature. *Reproductive Toxicology*, **42**:132–155.

Sathyanarayana S. (2008). Phthalates and children's health. *Current Problems in Pediatric and Adolescent Health Care*. **38**(2), 34–49.

Silvestre C., Duraccio D. and Cimmino S. (2011). Food packaging based on polymer nanomaterials. *Progress in Polymer Science*, **36**(12), 1766–1782.

Swan S. H. (2008). Environmental phthalate exposure in relation to reproductive outcomes and other health endpoints in humans. *Environmental Research*, **108**(2), 177–184.

Talsness C. E., Andrade A. J., Kuriyama S. N., Taylor J. A. and vom Saal F. S. (2009). Components of plastic: experimental studies in animals and relevance for human health. *Philosophical transactions of the Royal Society of London B Biological Sciences*, **364**(1526), 2079–2096.

Testai E., Hartemann P., Rastogi S. C., Bernauer U., Piersma A., De Jong W., Gulliksson H., Sharpe R., Schubert D. and Rodríguez-Farre E. (2016). The safety of medical devices containing DEHP plasticized PVC or other plasticizers on neonates and other groups possibly at risk (2015 update). *Regulatory Toxicology and Pharmacology*, **76**, 209–210.

Tyl R. W. (2014). Abbreviated assessment of bisphenol A toxicology literature. *Seminars in Fetal & Neonatal Medicine*, **19**(3), 195–202.

Vandenberg L. N., Maffini M. V., Sonnenschein C., Rubin B. S. and Soto A. M. 2009. Bisphenol-A and the great divide: a review of controversies in the field of endocrine disruption. *Endocrine Reviews*, **30**(1), 75–95.

Vogel S. A. (2009). The politics of plastics: the making and unmaking of bisphenol a "safety". *American Journal of Public Health*, **99**(3), S559–S566.

Völkel W., Colnot T., Csanády G. A., Filser J. G. and Dekant W. (2002). Metabolism and kinetics of bisphenol A in humans at low doses following oral administration. *Chemical Research in Toxicology*, **15**(10), 1281–1287.

vom Saal F. S. and Hughes C. (2005). An extensive new literature concerning low-dose effects of bisphenol A shows the need for a new risk assessment. *Environmental Health Perspectives*, **113**(8), 926–933.

Wolff M. S., Teitelbaum S. L., Pinney S. M., Windham G., Liao L., Biro F., Kushi L. H., Erdmann C., Hiatt R. A., Rybak M. E. and Calafat A. M. (2010). Investigation of relationships between urinary biomarkers of phytoestrogens, phthalates, and phenols and pubertal stages in girls. *Environmental Health Perspectives*, **118**(7), 1039–1046.

Wright S. L. and Kelly F. J. (2017). Plastic and human health: a micro issue? *Environmental Science & Technology*, **51**(12), 6634–6647.

Yang D., Shi H., Li L., Li J., Jabeen K. and Kolandhasamy P. (2015). Microplastic pollution in table salts from China. *Environmental Science & Technology*, **49**(22), 13622–13627.

Chapter 13

The need for a global plastic strategy

S. Kordella[1], H. K. Karapanagioti[2] and G. Papatheodorou[1]

[1]Department of Geology, University of Patras, Lab. of Marine Geology and Physical Oceanography, Greece
[2]Department of Chemistry, University of Patras, Greece

Keywords: Marine litter, Plastic litter, Plastic pollution, Plastic strategy, Policy, Single-use plastics

13.1 THE ENVIRONMENTAL PROBLEM

Plastic pollution in freshwater and marine areas has been widely recognized as one of the most crucial global concerns of our time. It has been estimated that in the last six decades, 8,300 million metric tonnes (Megatonnes; Mt) of plastic has been produced, most of it for the creation of disposable products. Of this, 6,300 Mt of plastic became waste, around 9% of which has been recycled, 12% incinerated and 79% accumulated in landfills or disposed in the natural environment (Geyer *et al.*, 2017) and the world's oceans: the final sink (Pham *et al.*, 2014; Ryan, 2015), causing environmental, economic, health and aesthetic implications (Engler, 2012; Rochman *et al.*, 2013a, b; Sheavly & Register, 2007; Silva-Iñiguez & Fischer, 2003).

Barring some local fluctuations, marine litter sources can be separated into sea-based (considered to contribute around 20% in total marine litter pollution) and land-based sources (which account for 80%) (UNEP, 2006). Sea-based

doi: 10.2166/9781789061697_0191

marine litter originates from fisheries and aquaculture, shipping (transport, military and tourism), offshore oil and gas exploration, and from illegal dumping at sea, etc., whilst, millions of tonnes of litter enter the marine environment from land-based sources through floodwaters and discharges from storm water drains, through rivers, streams, the littering of beaches and coastal areas, industrial facilities, landfills and illegal waste dumps located in proximity to the coast and/or water bodies, and via untreated municipal sewage.

Plastic is the most abundant litter material, accounting for 60–95% of marine litter pollution worldwide (Derraik, 2002; Galgani *et al.*, 2015) and sometimes accounting for up to 100% of floating litter (Galgani *et al.*, 2015). It was recently estimated that 8 Mt of plastic ends up in oceans every year (Jambeck *et al.*, 2015) and more than 5 trillion (5×10^{12}) plastic pieces weighing over 250,000 tonnes are afloat at sea (Eriksen *et al.*, 2014), whilst plastic particles and fibers are found today in tap water, beer and salt (Kosuth *et al.*, 2018; Karami *et al.*, 2017) with yet unknown impacts on public health. Plastic is persistent and lightweight. These two properties that make it so popular as a material are also the reasons that make it a threat to marine ecosystems and to wildlife. If their shape and size allows it (e.g., in the case of plastic cotton swabs), plastic litter is easily blown by high winds or drifted by heavy rainfall to waterways, from streets and poorly designed bins in population centers, from landfills and waste dumps, tourist beaches or through sewage water and from wastewater treatment plants (Mourgkogiannis *et al.*, 2018). Microplastics (GESAMP, 2015) and nanoplastics (Rios Mendoza *et al.*, 2018) – either as the product of fragmentation resulting from the exposure of macroplastics to the marine environment or directly produced – reach the marine environment through sewage waters (e.g., microfibers and microbeads from cosmetics) and runoffs (e.g., pellets), and accumulate with a disturbingly accelerating pace (Figure 13.1).

The ten most commonly found single-use plastic items in the marine environment together with lost and abandoned fishing gear account for at least 70% of total marine litter (Cau *et al.*, 2018; Fortibuoni *et al.*, 2019; Galgani *et al.*, 2015; Koutsodentris *et al.*, 2008; Thiel *et al.*, 2013; Topçu *et al.*, 2013). These items include plastic bags, water bottles, plastic cups, plastic cutlery, straws, etc.

The impact of marine litter on coastal and marine ecosystems and marine wildlife is reflected worldwide in the literature (Bernardini *et al.*, 2018; Green *et al.*, 2015; Green, 2016; Mordecai *et al.*, 2011; Panti *et al.*, 2019; Rochman, *et al.*, 2015). Impacts include the risk of entanglement and ingestion of litter by marine animals and birds (Bjorndal *et al.*, 1994; Campani *et al.*, 2013; De Pierrepont *et al.*, 2005; Tourinho *et al.*, 2010; Wilcox *et al.*, 2016), whilst microplastics and the fragmentation of plastics in the marine environment are a special risk to marine life (Gregory, 2009; Rochman *et al.*, 2013b). Microplastics and nanoplastics attract persistent organic pollutants (POPs) (Karapanagioti & Klontza, 2008; Takada & Karapanagioti, 2019), reaching up to one million times

Figure 13.1 Beach stranded microplastics produced from fragmentation due to the exposure of macroplastics to the marine environment and plastic pellets were among the various items of plastic litter found during beach litter monitoring surveys for the LIFE DEBAG project. (Photo: Stavroula Kordella, 2018).

higher concentrations than seawater (Rios Mendoza *et al.*, 2018). According to one study, it has been estimated that European citizens ingest up to 11,000 plastic fragments per year with their seafood (Van Cauwenberghe & Janssen, 2014). However, little is yet known relating to the impact this has on human health.

If current plastic production coupled with waste management trends continue, roughly 12,000 Mt of plastic waste will be disposed of in landfills or in the natural environment by 2050 (Geyer *et al.*, 2017). Remedial actions to improve plastic pollution, such as cleaning, have been tried and have been found to be insufficient and cost ineffective. These facts emphasize the important need for drastic preventive action aiming at the sources of marine plastic litter pollution (UNEP, 2009). Such actions include comprehensive, binding, global strategies and policies. A lot of effort has been made towards the development and implementation of such strategies and policies around the globe. In this chapter, existing actions are explored and the gaps to be filled that may pave the way for successful implementation, judged by results, are identified.

13.2 REVIEW OF KEY STRATEGIES AND POLICIES

Looking at the current status of plastic marine litter pollution, it is self-evident that there is an urgent need to react. The only known way to approach this issue is through development and implementation of strategies and policies reinforced by expanded information and education campaigns, and by the employment of green levies and economic incentives. Several countries have acted to confront the marine litter crisis but the situation is still not improving significantly at a global level (UNEP, 2009; Xanthos & Walker, 2017).

13.2.1 International strategies and policies on marine litter pollution

Strategies and policies on single-use plastics at a regional or country level, as levies or bans on single-use plastic bags may be increasing (Heidbreder *et al.*, 2019; Saidan *et al.*, 2017; Xanthos & Walker, 2017) but there are only a few international strategies and policies that address plastic, marine pollution directly. These amount to four main international strategies and policies: MARPOL, the Honolulu Strategy, the Global Partnership on Marine Litter, and the United Nations Environmental Program (UNEP) Clean Seas campaign, which are each described below.

13.2.1.1 MARPOL 73/78

The *International Convention for the Prevention of Pollution from Ships* is the main international convention concerning prevention of pollution of the marine environment by ships from operational or accidental causes. MARPOL entered into force on 2 October 1983 and has been updated by amendments over the years.

Annex V, *Prevention of Pollution by Garbage from Ships*, entered into force on 31 December 1988 and deals with different types of garbage, and specifies the distances from land and the ways in which they may be disposed. The most important trait of the Annex is the complete ban that is imposed on the disposal into the sea of all forms of plastics (IMO, 2019).

Even though, since January 2018, 156 states and member nations have been subject to MARPOL requirements, regardless of where they sail, research has shown that marine litter has increasingly risen since MARPOL 73/78 was signed (Borrelle *et al.*, 2017; Jambeck *et al.*, 2015; Koutsodentris *et al.*, 2007; Xanthos & Walker, 2017). The deterioration of the marine environment stems from the fact that MARPOL Annex V, the oldest strategy directly concerning marine litter, is restricted to litter originating from ships (marine-based) which accounts for less than 20% of the total marine litter pollution (as fishery related litter is also classified as marine-based), while (as noted above) the large majority (80%) of marine litter originates from land-based sources (UNEP, 2006).

13.2.1.2 The Honolulu Strategy

The *Honolulu Strategy* is a framework document created the National Oceanic and Atmospheric Administration (NOAA) and United Nations Environment Programme (UNEP) in 2011, concerning a global, comprehensive effort to reduce marine litter and its ecological, public health and economic impacts. The aim of this document is to help improve collaboration among groups and countries from around the globe concerning marine litter pollution, and to serve as a framework as well as a tool to develop and monitor marine litter projects.

The Honolulu Strategy is intended for use as:

- a planning tool for developing spatially or sector-specific marine litter projects;
- a common frame of reference for collaboration and sharing of best practices and lessons learned;
- a monitoring tool to measure progress across multiple programs and projects (UNEP & NOAA, 2015).

Two parts of the Honolulu Strategy are highly important: the part that focuses on market-based instruments (e.g., levies on bags) for minimizing waste; and the part that concerns the development of policies and regulations to reduce marine litter (e.g., bans on plastic bags and microbeads in cosmetics) (Xanthos & Walker, 2017).

13.2.1.3 The global partnership on marine litter

In June 2012, at the Rio+20 conference in Brazil, the Global Partnership on Marine Litter (GPML) was launched. The GPML is a voluntary, open-ended partnership engaging international agencies, governments, businesses, academia, local authorities and NGOs. It is hosted by UNEP and aims to achieve a significant reduction of marine litter, with a deadline of 2025. The GPML seeks to protect public health and the global environment by the reduction and management of marine litter through the following specific objectives:

- to enhance international cooperation and coordination through the promotion and implementation of the Honolulu Strategy (see 13.2.1.2) as well as the Honolulu Commitment (a multi-stakeholder pledge);
- to promote knowledge, management, information sharing and monitoring of progress on the implementation of the Honolulu Strategy;
- to promote resource efficiency and economic development through waste prevention e.g., by promoting the 4Rs (reduce, re-use, recycle and re-design) and by recovering valuable materials and/or energy from waste;
- to increase awareness of the sources and fate and impacts of marine litter; and
- to assess emerging issues related to the fate and implications of marine litter, including (micro)plastics uptake in the food web and the associated transfer of pollutants and impacts.

13.2.1.4 UNEP clean seas campaign

In February 2017, UNEP launched the Clean Seas campaign, engaging governments, the general public and the private sector, and persuading them to voluntarily commit to actions to reduce plastic pollution. Fifty governments – accounting for more than half the world's coastlines – have signed up to the Clean Seas campaign with many making specific commitments to protect the oceans, encourage recycling and cut back on single-use plastics. The campaign contributes to the goals of the Global Partnership on Marine Litter.

Pledges that have been made in the framework of the UNEP Clean Seas campaign include:

- Belgium, Brazil, the Dominican Republic, Panama and the Philippines are developing and/or adopting national plans and legislation to combat marine litter;
- Canada, which is the country with the world's longest coastline, is funding community-based programmes such as beach clean-ups and continuing research into the impact of microplastics. It is also developing regulations to ban the manufacture and sale of cosmetics containing microbeads;
- Indonesia has committed to reduce plastic waste by 70% by 2030;
- Kenya, Jordan, Madagascar and Chile have banned or pledged to ban single-use or non-biodegradable plastic bags;
- Nigeria, one of the world's top 10 plastic polluters, has pledged to open 26 plastic recycling plants;
- Denmark, Finland, Iceland and Sweden have committed to implement the "Nordic programme" on a sustainable approach to plastics by preventing plastic waste, encouraging recycling and promoting a circular economy;
- New Zealand prohibited the sale and manufacture of wash-off products that contain plastic microbeads from 7 June 2018 and the Ministry of Environment has confirmed that single-use plastic shopping bags up to 70 microns in thickness will be phased out with regulations to come into force from 1 July 2019.

The Clean Seas campaign monitors these pledges and aims to get more countries to commit to action. It also aims for the increasing cooperation of businesses. To date, many European retailers have committed to plastic-free aisles and products while some restaurants have pledged to phase out plastic straws.

13.2.1.5 The Basel convention

The Basel Convention on the Control of Transboundary Movements of Hazardous Wastes and their Disposal entered into force on 5 May 1992 and has 187 parties (countries from around the world). The Convention aims to minimize the generation of hazardous wastes and "other wastes" (i.e. household waste and incinerator ash), to control their transboundary movements, and promote their environmentally sound management.

Under the Convention, some plastics are listed as "hazardous wastes" and household wastes may also include plastics. Therefore the provisions of the Convention already applied to plastic waste but, during the recent Basel Conference of the Parties (held from 29 April–10 May 2019), a major step forward was taken when the Convention was amended to include plastic waste in a legally-binding framework which "will make global trade in plastic waste more transparent and better regulated, whilst also ensuring that its management is safer for human health and the environment" (Secretariat of the Basel, Rotterdam and Stockholm Conventions, 2019). On the basis of transboundary movements of hazardous waste, plastic considered contaminated, mixed and unsuitable for recycling will be subject to controls and the consent of importing countries will be required, thus promoting recycling within the boundaries of the countries that are major plastic waste generators and providing an important tool for developing countries to deny unrecyclable plastic.

13.2.2 European strategies and policies on marine litter pollution

13.2.2.1 The marine strategy framework directive

Based on its adverse effects, marine litter pollution has been included in the 11 qualitative descriptors set by the European Marine Strategy Framework Directive (MSFD) (Directive 2008/56/EC) (European Parliament, 2008; Galgani *et al.*, 2013a; Galgani *et al.*, 2010). The MSFD requires each descriptor to maintain or reach Good Environmental Status (GES) for all European marine waters by 2020 at the latest (Article 1). Concerning Descriptor 10 (marine litter), the MSFD requires EU Member States to ensure that, by 2020, "properties and quantities of marine litter do not cause harm to the coastal and marine environment". At EU level, the above-mentioned MSFD is the dedicated binding legal instrument for assessing, monitoring, setting targets and reaching GES about marine litter; appointed by the Member States to support them in reaching GES for marine litter, it is co-chaired by the Joint Research Center (JRC) and has developed, inter alia, 'Guidance on Monitoring of Marine Litter in the European Seas' (Galgani *et al.*, 2013b). The last report published by the JRC on the top 10 litter items most frequently found on European beaches reflects monitoring results from the EU Member States and the Regional Seas Conventions and the analysis which was the basis for the Commission proposal on single-use plastic items (see 13.2.2.3 below). Through the adoption of the MSFD in 2008, the EU established a framework to protect and sustainably use its seas and oceans, requiring implementation of marine strategies from EU member countries.

13.2.2.2 EU Directive 2015/720 on plastic bags

On 29 April 2015, the European Parliament introduced Directive 2015/720/EC for the reduction on the consumption of lightweight (15–50 microns wall thickness) plastic carrier bags, many of which end up as waste in the marine environment (European Parliament, 2015). Lightweight plastic bags represent most of the total number of plastic carrier bags in the EU and are less frequently reused than thicker plastic carrier bags. Consequently, lightweight plastic carrier bags become waste more quickly and are more prone to littering due to their light weight. Plastic bags account for almost 5% of litter found on beaches but their abundance rises to a 30% of the litter found on the seabed around the European coastline (Galgani *et al.*, 1995, 2000; Ramirez-Llodra *et al.*, 2013). Member States were bound to take measures to achieve a sustained reduction in the consumption of lightweight plastic carrier bags in their countries.

The measures taken by Member States should include either or both of the following: (a) the adoption of measures ensuring that the annual consumption level does not exceed 90 lightweight plastic carrier bags per capita by 31 December 2019 and 40 lightweight plastic carrier bags per capita by 31 December 2025; and (b) the adoption of instruments ensuring that, by 31 December 2018, lightweight plastic carrier bags are not provided free of charge at the point of sale of goods or products, unless equally effective instruments are implemented. The Directive gave the option of exclusion from these measures of very lightweight plastic carrier bags (wall thickness below 15 microns) provided as primary packaging for loose food when required for hygiene purposes or when their use helps prevent food wastage.

As a complement to EU preventive measures and strategies, EU funds (including from the EU LIFE programme) support action to assist the EU in fighting for the preservation of the environment and the effective implementation of EU policy. Concerning marine litter, LIFE projects have helped implement EU policy in areas such as the circular economy, single-use plastics, etc., and have involved awareness campaigns and clean-up operations on beaches or at sea, with the active participation of citizens, fishermen, businesses and other stakeholders. One example is the LIFE DEBAG project (LIFE14 GIE/GR/001127) which implemented an integrated information and awareness-raising campaign for the reduction of plastic bags in the marine environment of Greece, both at local and national levels. The LIFE DEBAG project contributed heavily to the integration of EU Directive 2015/720 into Greek legislation through recommendations developed within a series of consultation fora. The Greek legislation imposed a green levy on lightweight plastic bags from 1 January 2018 for the first time, leading to a 60–80% decrease in plastic bag consumption across the country, after one year of enforcement of the law. For the implementation of a local and more intensive information campaign, the island of Syros, located in the Aegean Sea, was selected as a pilot area. A reduction in the accumulation of plastic bags of 85% on beaches and 60% on the seafloor around Syros had occurred by the end of the project, which was induced directly from the LIFE DEBAG intensive awareness campaign on the island, a fact that was made evident through an exhaustive monitoring of Syros's marine environment before and during the realization of the campaign. The positive results on the marine environment of the pilot area of the campaign proved that information campaigns can be effective. This is true only if free reusable alternatives are provided, if all relevant stakeholders in decision-making processes are involved, and if the campaign impact is exhaustively monitored before, during and after the activities. The participation of citizens and the communication of the results to the public as a positive feedback of changing consumer habits and positive reinforcement were the key elements of this campaign (EU DG Environment, 2018).

Monitoring of a strategy's impact on the marine environment is crucial and really proves the effectiveness of measures taken. In the case of the UK, Maes *et al.*, (2018) estimated a 30% decrease in plastic bags on the seabed around the UK's coasts in the same period as green levies were introduced in European countries, highlighting the effectiveness of economic disincentives as tools for reductions in numbers of single-use plastic bags.

13.2.2.3 The EU strategy for plastics in a circular economy

Pollution of the seas from plastics and microplastics is one of the three major areas of the Strategy for Plastics adopted by the European Commission on 16 January 2018 (European Commission, 2018a). Most of the proposed actions are directly or indirectly related to marine litter.

According to the EU Strategy for Plastics, "by 2030, all plastics packaging placed on the EU market is either reusable or can be recycled in a cost-effective manner" (European Commission, 2018a). Consumption of single-use plastics, including over-packaging, will be reduced and the intentional use of microplastics will be restricted.

The Commission will revise the legislative requirements for placing packaging on the market focusing on defining the concept of design for recyclability. The goal is to decrease the quantity of waste generated and to avoid packaging materials ending up as litter, being incinerated or landfilled instead of recycled. The Commission invited the plastics industry to take an active part in the process by supporting innovation in this domain through a pledging exercise which is being launched by the strategy which aims to see 10 Mt of recycled plastics in new products by 2025.

The Commission will propose harmonised rules for defining and labelling compostable and biodegradable plastics, which may be used as an alternative to conventional plastics, but the lack of clear labelling and waste collection and treatment may lead to plastics leakage, The Commission aims to reduce discharges of waste from ships at sea by ensuring the availability of adequate port reception facilities and the use of those facilities by ships visiting EU ports through the new Port Reception Facilities Directive (European Commission, 2018b). The Directive, which was proposed on 16 January 2018, is based on the international obligations in the MARPOL Convention (see 13.2.1.1 above).

The new Plastics Strategy of the European Union aims to curb plastic waste by targeting single-use plastics and fishing gear, supporting national awareness campaigns and determining the scope of new EU-wide rules that were proposed in 2018 and constitute the Single-Use Plastics Directive (see 13.2.2.4 below) based on stakeholder consultation and evidence.

Finally, the Commission has started work to restrict the use of microplastics that are intentionally added to products through the Registration, Evaluation, Authorisation, and Restriction of Chemicals (REACH) regulation (European Commission, 2018a). Regarding unintentional release of microplastics, the Commission is examining options such as labelling, minimum requirements for product design and durability, methods to assess quantities and pathways of microplastics in the environment, and funding for targeted research and innovation.

13.2.2.4 The single-use plastics directive

On 19 December 2018, the European Parliament and the Council of the European Union reached a provisional political agreement on the ambitious new measures (European Commission, 2018c) proposed by the European Commission to tackle marine litter at its source, targeting the ten plastic products most often found on EU beaches as well as abandoned fishing gear, which account for at least 70% of total marine litter (Figure 13.2), and oxo-degradable plastics.

These measures, which are part of the EU Strategy for Plastics (see 13.2.2.3 above), were adopted by the European Parliament on 27 March 2019, constituting the new EU Directive on Single-Use Plastics: the most ambitious legal instrument at a global level addressing marine litter. It envisages different measures to apply to different product categories. Where alternatives are easily available and affordable, single-use plastic products will be banned from the market (such as plastic cotton buds, cutlery, plates, straws, drink stirrers, sticks for balloons, products made of oxo-degradable plastic, and food and beverage containers made of expanded polystyrene). For other products, the focus is on limiting their use and/or abundance in the marine environment through:

- setting national reduction targets for consumption (of food containers and drinks cups);
- enforcing design and labelling requirements (sanitary towels, wet wipes, balloons);
- establishing waste management/clean-up obligations for producers (food containers, crisps and sweets packaging, drinks containers, cigarette butts, wet wipes, balloons and lightweight plastic bags);

Figure 13.2 Classification of beach stranded litter for the LIFE DEBAG project showing the abundance of single-use litter such as water bottle caps and straws etc., together with fishery related litter, which accounts for more than 70% of the total marine litter on European coasts. (Photo: Stavroula Kordella, 2019).

- encouraging collection, e.g. through deposit refund schemes, of 90% of single-use plastic drinks bottles by 2025;
- introducing awareness-raising measures about the negative impact of single-use plastics and fishing gear, and the re-use systems and waste management options for these products.

13.2.2.5 The Barcelona convention and its protocols

The Convention for the Protection of the Mediterranean Sea Against Pollution ('MAP', or 'Barcelona Convention') was adopted on 16 February 1976 by the Conference of Plenipotentiaries of the Coastal States of the Mediterranean Region for the Protection of the Mediterranean Sea, held in Barcelona, in conjunction with two Protocols addressing the prevention of pollution by dumping from ships and aircraft and cooperation in combating pollution in cases of emergency.

Seven Protocols addressing specific aspects of Mediterranean environmental conservation complete the MAP legal framework but those related to marine litter are:

- The Dumping Protocol: The Protocol for the Prevention of Pollution in the Mediterranean Sea by Dumping from Ships and Aircraft (adopted in 1976); and

- The Land-Based Sources (LBS) Protocol: The Protocol for the Protection of the Mediterranean Sea against Pollution from LBS and Activities (adopted in 1980).

In the framework of the LBS Protocol, a marine litter management strategy was developed in 2012 which was accompanied by the Regional Action Plan on Marine Litter. The Action Plan was adopted during the 18th Conference of the Contracting Parties to the Barcelona Convention and its Protocols held in Istanbul in 2013 and entered into force – and thus became legally binding – on 8 July 2014. Its aim is to reduce the impact of marine litter on the environment, human health and on the Mediterranean economy by enhancing cooperation, promoting and implementing international and regional marine waste initiatives, and raising awareness and knowledge.

13.2.3 National and local initiatives

More than 60 countries around the globe have taken measures on plastic bags, microbeads, products, straws and plastic cutlery and plastic cotton swabs, and the number is constantly rising. Policies against microbeads and other single-use plastics are more recent (2014 in the case of microbeads) but initiatives for plastic bags began in 1991 (Xanthos & Walker, 2017). According to a UNEP (2018) report that analysed over 140 regulations at national and local levels concerning bans and levies on plastic bags, there is not enough information to draw safe conclusions on their environmental impacts, although in 30% of the cases there was a reduction in plastic bags either in the consumption or in the marine environment. In 50% of the investigated cases, there is no information on impact, partially due to lack of monitoring and reporting, and partially because many of the measures analysed have been implemented recently (UNEP, 2018). In 20% of the cases of countries that have banned plastic bags, no to little impact was reported; this was attributed to lack of enforcement and lack of affordable alternatives (UNEP, 2018).

13.3 CONCLUSIONS

Plastic marine pollution is an international issue, which knows no boundaries (Politikos *et al.*, 2017; Villarrubia-Gómeza *et al.*, 2018). Therefore, in order to tackle this fast growing problem, there is a need for a global governance approach (Vince & Stoett, 2018). Countries should join to set reduction targets and develop policy schemes combined with exhaustive monitoring both in the marine environment and in the consumption of plastic packaging, single-use products and products that generate microplastics, etc. There are many regional/national policies that concern single-use plastic production prevention and plastic pollution mitigation but fewer international policies, none of which have yet achieved results that compensate for the global enormity and accelerating growth of the problem (Borrelle *et al.*, 2017; UN Environment, 2017).

Binding agreements for countries and industry (Borrelle *et al.*, 2017), integrated global strategies (Dauvergne, 2018) that include actions on solid waste management, collection and treatment of wastewater and rain-storm water, extended producer responsibility for single-use plastics and packaging, economic incentives or disincentives for citizens and industry, and public awareness-raising campaigns combined with affordable reusable alternatives can all boost implementation levels and mitigate the plastic pollution problem dramatically.

Although existing international policies and strategies acknowledge marine litter as a global, multi-parametric threat, they lack binding commitments for countries as well as monitoring schemes to evaluate their effectiveness and measure their impact not only on plastic product consumption but most importantly on the marine environment itself. Therefore, the solution may lie within a global strategy containing three axes: policies, awareness-raising and marine litter monitoring, with binding targets that should come in effect shortly, as the magnitude and rapidity of the issue calls for immediate action.

REFERENCES

Bernardini I., Garibaldi F., Canesi L., Fossi M. C. and Baini M. (2018). First data on plastic ingestion by blue sharks (Prionaceglauca) from the Ligurian Sea (North-Western Mediterranean Sea). *Marine Pollution Bulletin*, **135**, 303–310.

Bjorndal K. A., Bolten A. B. and Lagueux C. J. (1994). Ingestion of marine debris by juvenile sea turtles in coastal Florida habitats. *Marine Pollution Bulletin*, **28**(3), 154–158.

Borrelle S. B., Rochman C. M., Liboiron M., Bond A. L., Lusher A., Bradshaw H. and Provencher J. F. (2017). Opinion: Why we need an international agreement on marine plastic pollution. *Proceedings of the National Academy of Sciences of the United States of America*, **114**(38), 9994–9997.

Campani T., Baini M., Giannetti M., Cancelli F., Mancusi C., Serena F., Marsili L., Casini S. and Fossi M. C. (2013). Presence of plastic debris in loggerhead turtle stranded along the Tuscanycoasts of the Pelagos Sanctuary for Mediterranean Marine Mammals (Italy). *Marine Pollution Bulletin*, **74**, 225–230.

Cau A., Bellodi A., Moccia D., Mulas A., Pesci P., Cannas R., Pusceddu A. and Follesa M. C. (2018). Dumping to the abyss: single-use marine litter invading bathyal plains of the Sardinian margin (Tyrrhenian Sea). *Marine Pollution Bulletin*, **135**, 845–851.

Dauvergne P. (2018). Why is the global governance of plastic failing the oceans? (2018). *Global Environmental Change*, **51**, 22–31.

De Pierrepont J. F., Dubois B., Desormonts S., Santos M. B. and Robin J. P. (2005). Stomach contents of English Channel cetaceans stranded on the coast of Normandy. *Journal of the Marine Biological Association of the United Kingdom*, **85**, 1539–1546.

Derraik J. G. B. (2002). The pollution of the marine environment by plastic debris: a review. *Marine Pollution Bulletin*, **44**, 842–852.

Engler R. E. (2012). The complex interaction between marine debris and toxic chemicals in the ocean. *Environmental Science and Technology*, **46**(22), 12302–12315.

Eriksen M., Lebreton L. C. M., Carson H. S., Thiel M., Moore C. J., Borerro J. C., Galgani F., Ryan P. G. and Reisser J. (2014). Plastic pollution in the world's oceans: more than 5 trillion plastic pieces weighing over 250,000 tons afloat at sea. *PLoS ONE*, **9**(12), e111913.

European Commission (2018a). Communication from the Commission to the European Parliament, the Council, the European Economic and Social Committee and the Committee of the Regions. A European Strategy for Plastics in a Circular Economy. COM/2018/028 final. See: https://eur-lex.europa.eu/legal-content/EN/TXT/HTML/?uri=CELEX:52018DC0028&from=EN (accessed 10 February 2019).

European Commission (2018b). Proposal for a DIRECTIVE OF THE EUROPEAN PARLIAMENT AND OF THE COUNCIL on Port Reception Facilities for the Delivery of Waste from Ships, Repealing Directive 2000/59/EC and Amending Directive 2009/16/EC and Directive 2010/65/EU. COM/2018/033 final – 2018/012 (COD). See: https://eur-lex.europa.eu/legal-content/EN/TXT/?uri=CELEX: 52018PC0033 (accessed 15 May 2019).

European Commission (2018c). Proposal for a Directive of the European Parliament and of the Council on the reduction of the impact of certain plastic products on the environment. COM/2018/340 final – 2018/0172 (COD). See: https://eur-lex.europa.eu/legal-content/en/ALL/?uri=CELEX%3A52018PC0340 (accessed 15 May 2019).

European Commission Directorate General for Environment (EU DG Environment) (2018). LIFE and the EU Plastics Strategy. See: https://ec.europa.eu/easme/sites/easme-site/files/life_plastics_web.pdf?pk_campaign=LIFE+Newsletter+Jan-Feb2019&fbclid=IwAR08fXcsUUJzV0cEfJA1 FdgqAiWK52NzV6OE9DcuzjfP_U-YIrNIJhwflOQ (accessed 20 February 2019), p. 24.

European Parliament (2008). Directive 2008/56/EC of the European Parliament and of the Council of 17 June 2008 establishing a framework for community action in the field of marine environmental policy (Marine Strategy Framework Directive). *Official Journal of the European Union*. OJ L 164, 25.6.2008, 19–40. See: http://data.europa.eu/eli/dir/2008/56/oj (accessed 15 May 2019).

European Parliament (2015). Directive (EU) 2015/720 of the European Parliament and of the Council of 29 April 2015 amending Directive 94/62/EC as regards reducing the consumption of lightweight plastic carrier bags. *Official Journal of the European Union*. L 115, 6.5.2015, 11–15. See: https://publications.europa.eu/en/publication-detail/-/publication/58d93aee-f3bc-11e4-a3bf-01aa75ed71a1/language-en/format-PDF/source-87251879 (accessed 7 January 2019).

Fortibuoni T., Ronchi F., Mačić V., Mandić M., Mazziotti C., Peterlin M., Prevenios M., Prvan M., Somarakis S., Tutman P., BojanićVarezić D., KovacVirsek M., Vlachogianni T. and Zeri C. (2019). A harmonized and coordinated assessment of the abundance and composition of seafloor litter in the Adriatic-Ionian macroregion (Mediterranean Sea). *Marine Pollution Bulletin*, **139**, 412–426.

Galgani F., Burgeot T., Bocquéné G., Vincent F., Leauté J. P., Labastie J., Forest A. and Guichet R. (1995). Distribution and abundance of debris on the continental shelf of the North-Western Mediterranean Sea. *Marine Pollution Bulletin*, **30**, 58–62.

Galgani F., Leaute J. P., Moguedet P., Souplet A., Verin Y., Carpentier A., Goraguer H., Latrouite D., Andral B., Cadiou Y., Mahe J. C., Poulard J. C. and Nerisson P.

(2000). Litter on the sea floor along European coasts. *Marine Pollution Bulletin*, **40**, 516–527.

Galgani F., Fleet D., Van Franeker J., Katsanevakis S., Maes T., Mouat J., Oosterbaan L., Poitou I., Hanke G., Thompson R., Amato E., Birkun A. and Jansse E. (2010). Marine Strategy Framework Directive Task Group 10 Report on Marine Litter European Union, IFREMER and ICES. See: http://ec.europa.eu/environment/marine/pdf/9-Task-Group-10.pdf (accessed 10 January 2019).

Galgani F., Hanke G., Werner S. and De Vrees L. (2013a). Marine litter within the European Marine Strategy Framework Directive. *ICES Journal of Marine Science*, **70**, 1055–1064.

Galgani F., Hanke G., Werner S., Oosterbaan L., Nilsson P., Fleet D., Kinsey S., Thompson R. C., Van Franeker J., Vlachogianni T., Scoullos M., Mira Veiga J., Palatinus A., Matiddi M., Maes T., Korpinen S., Budziak A., Leslie H., Gago J. and Liebezeit G. (2013b). Guidance on Monitoring of Marine Litter in European Seas. MSFD GES Technical Subgroup on Marine Litter (TSG-ML). Joint Research Centre of the European Commission, Report EUR 26113 EN. Publications Office of the European Union, Luxembourg, 120p.

Galgani F., Hanke G. and Maes T. (2015). Global distribution, composition and abundance of marine litter. In: Marine Anthropogenic Litter, M. Bergmann, L. Gutow and M. Klages (eds.), Springer, Cham.

GESAMP (2015). Chapter 3.1.2 Defining 'microplastics'. Sources, fate and effects of microplastics in the marine environment: a global assessment. In: (IMO/FAO/UNESCO–IOC/UNIDO/WMO/IAEA/UN/UNEP/UNDP Joint Group of Experts on the Scientific Aspects of Marine Environmental Protection (GESAMP)), P. J. Kershaw (ed.), Rep. Stud. GESAMP No. 90, 96p, London.

Geyer R., Jambeck J. R. and Law K. L. (2017). Production, use, and fate of all plastics ever made. *Science Advances*, **3**, 1700782.

Green D. S. (2016). Effects of microplastics on European flat oysters, *Ostrea edulis* and their associated benthic communities. *Environmental Pollution*, **216**, 95–103.

Green D. S., Boots B., Blockley D. J., Rocha C. and Thompson R. (2015). Impacts of discarded plastic bags on marine assemblages and ecosystem functioning. *Environ. Sci. Technol.*, **49**(9), 5380–5389.

Gregory M. R. (2009). Environmental implications of plastic debris in marine settings–entanglement, ingestion, smothering, hangers on, hitch-hiking and alien invasions. *Philosophical Transactions of the Royal Society of London. B, Biological Sciences*, **364**(1526): 2013–2025.

Heidbreder L. M., Bablok I., Drews S. and Menzel C. (2019). Tackling the plastic problem: A review on perceptions, behaviors, and interventions. *Science of the Total Environment*, **668**, 1077–1093.

International Maritime Organization (IMO) (2019). International Convention for the Prevention of Pollution from Ships (MARPOL). See: http://www.imo.org/en/about/conventions/listofconventions/pages/international-convention-for-the-prevention-of-pollution-from-ships-(marpol).aspx (accessed 5 February 2019).

Jambeck J. R., Geyer R., Wilcox C., Siegler T. R., Perryman M., Andrady A., Narayan R. and Law K. L. (2015). Plastic waste inputs from land into the ocean. *Science*, **347**(6223), 768–771.

Karami A., Golieskardi A., Keong Choo C., Larat V., Galloway T. S. and Salamatinia B. (2017). The presence of microplastics in commercial salts from different countries. *Scientific Reports*, **7**, e46173.

Karapanagioti H. K. and Klontza I. (2008). Testing phenanthrene distribution properties of virgin plasticpellets and plastic eroded pellets found on Lesvosisland beaches (Greece). *Marine Environmental Research*, **65**, 283–290.

Kosuth M., Mason S. A. and Wattenberg E. V. (2018). Anthropogenic contamination of tap water, beer, and sea salt. *PLoS ONE*, **13**(4), e0194970.

Koutsodendris A., Papatheodorou G., Kougiourouki O. and Georgiadis M., 2008. Benthic marine litter in four Gulfs in Greece, Eastern Mediterranean; abundance, composition and source identification. *Estuarine, Coastal and Shelf Science*, **77**, 501–512.

Maes T., Barry J., Leslie H. A., Vethaak A. D., Nicolaus E. E. M., Law R. J., Lyons B. P., Martinez R., Harley B. and Thain J. E. (2018). Below the surface: twenty-five years of seafloor litter monitoring in coastal seas of North West Europe (1992–2017). *Science of the Total Environment*, **630**, 790–798.

Mordecai G., Tyler P. A., Masson D. G. and Huvenne V. A. I. (2011). Litter in submarine canyons off the west coast of Portugal. *Deep-Sea Research Part II: Topical Studies in Oceanography*, **58**, 2489–2496

Mourgkogiannis N., Kalavrouziotis I. K. and Karapanagioti H. K. (2018). Questionnaire-based survey to managers of 101 wastewater treatment plants in Greece confirms their potential as plastic marine litter sources. *Marine Pollution Bulletin*, **133**, 822–827.

Panti C., Baini M., Lusher A., Hernandez-Milan G., Bravo Rebolledo E. L., Unger B., Syberg K., Simmonds M. P. and Fossi M. C. (2019). Marine litter: one of the major threats for marine mammals. *Outcomes from the European Cetacean Society workshop. Environmental Pollution*, **247**, 72–79.

Pham C. K., Ramirez-Llodra E., Alt C. H. S., Amaro T., Bergmann M., Canals M., Company J. B., Davies J., Duineveld G., Galgani F., Howell K. L., Huvenne V. A. I., Isidro E., Jones D. O. B., Lastras G., Morato T., Gomes-Pereira J. N., Purser A., Stewart H., Tojeira I., Tubau X., Van Rooij D. and Tyler P. A. (2014). Marine litter distribution and density in European Seas, from the shelves to deep basins. *PLoS ONE*, **9**, e95839.

Politikos D. V., Ioakeimidis C., Papatheodorou G. and Tsiaras K. (2017). Modelling the fate and distribution of floating litter particles in the Aegean Sea (E. *Mediterranean). Frontiers in Marine Science*, **4**, 191.

Ramirez-Llodra E., De Mol B., Company J. B., Coll M. and Sardà F. (2013). Effects of natural and anthropogenic processes in the distribution of marine litter in the deep Mediterranean Sea. *Progress in Oceanography*, **118**, 273–287.

Rios Mendoza L. M., Karapanagioti H. K. and and RamírezÁlvarez N., (2018). Micro (nanoplastics) in the marine environment: Current knowledge and gaps. *Current Opinion in Environmental Science & Health*, **1**, 47–51.

Rochman C. M., Browne M. A., Halpern B. S., Hentschel B. T., Hoh E., Karapanagioti H. K. and Thompson R. C. (2013a). Policy: classify plastic waste as hazardous. *Nature*, **494** (7436), 169–171.

Rochman C. M., Hoh E., Kurobe T. and Teh S. (2013b). Ingested plastic transfers hazardous chemicals to fish and induces hepatic stress. *Scientific Reports*, **3**, 3263.

Rochman C. M., Tahir A., Williams S. L., Baxa D. V., Lam R., Miller J. T., Teh F. C., Werorilangi S. and Teh S. J. (2015). Anthropogenic debris in seafood: plastic debris and fibers from textiles in fish and bivalves sold for human consumption. *Scientific Reports*, **5**, e14340.

Ryan P. G. (2015). A brief history of marine litter research. In: Marine Anthropogenic Litter, M. Bergmann, L. Gutow and M. Klages (eds.), Springer, Cham.

Saidan M. N., Ansour L. M. and Saidan H., 2017. Management of plastic bags waste: an assessment of scenarios in Jordan. *Journal of Chemical Technology and Metallurgy*, **52**(1), 148–154.

Secretariat of the Basel, Rotterdam and Stockholm Conventions (2019). Governments Agree Landmark Decisions to Protect People and Planet from Hazardous Chemicals and Waste, Including Plastic Waste. See: http://www.brsmeas.org/?tabid=8005 (accessed 17 May 2019).

Sheavly S. B. and Register K. M. (2007). Marine Debris & Plastics: Impacts and Solutions. *Journal of Polymers and Environment*, **15**, 301–305.

Silva-Iñiguez L. and Fischer D. W. (2003). Quantification and classification of marine litter on the municipal beach of Ensenada, Baja California, Mexico. *Marine Pollution Bulletin*, **46**, 132–138.

Takada H. and Karapanagioti H. K. (eds.). (2019). Hazardous Chemicals Associated with Plastics in the Marine Environment, Handbook of Environmental Chemistry, No 78. Springer International Publishing AG, Switzerland.

Thiel M., Hinojosa I. A., Miranda L., Pantoja J. F., Rivadeneira M. M. and Vásquez N. (2013). Anthropogenic marine debris in the coastal environment: A multi-year comparison between coastal waters and local shores. *Marine Pollution Bulletin*, **71**, 307–316.

Topçu E. N., Tonay A. M., Dede A., Öztürk A. A. and Öztürk B. (2013). Origin and abundance of marine litter along sandy beaches of the Turkish Western Black Sea Coast. *Marine Environmental Research*, **85**, 21–28.

Tourinho P. S., Ivar do Soul J. A. and Fillmann G. (2010). Is marine debris ingestion still a problem for the coastal marine biota of southern Brazil? *Mar.Pollut. Bull.* **60**, 396–401.

UN Environment (2017). Combating Marine Plastic Litter and Microplastics: An Assessment of the Effectiveness of Relevant International, Regional and Subregional Governance Strategies and Approaches. See: https://papersmart.unon.org/resolution/ uploads/unea-3_mpl_assessment-2017oct05_unedited_adjusted.pdf (accessed 10 January 2019).

UNEP (United Nations Environment Programme) (2006). The State of the Marine Environment: A regional assessment. Global Programme of Action for the Protection of the Marine Environment from Land-based Activities. United Nations Environment Programme, The Hague.

UNEP (United Nations Environment Programme) (2009). Marine Litter: A Global Challenge. UNEP, Nairobi, 11pp.

UNEP (United Nations Environment Programme) (2018). Single-Use Plastics: A Roadmap for Sustainability. See: https://www.euractiv.com/wp-content/uploads/sites/2/2018/06/WED-REPORT-SINGLE-USE-PLASTICS.pdf (accessed 20 February 2018).

UNEP (United Nations Environment Programme) and NOAA (National Oceanic and Atmospheric Administration) (2015). The Honolulu Strategy, A Global Framework for Prevention and Management of Marine Debris. Report. See: https://marinedebris.noaa.gov/honolulu-strategy (accessed 9 January 2019).

Van Cauwenberghe L. and Janssen C. R. (2014). Microplastics in bivalves cultured for human consumption. *Environmental Pollution*, **193**, 65–70.

Villarrubia-Gómez P., Cornell S. and Fabres J. (2018). Marine plastic pollution as a planetary boundary threat – The drifting piece in the sustainability puzzle. *Marine Policy*, **96**, 213–220.

Vince J. and Stoett P. (2018). From problem to crisis to interdisciplinary solutions: Plastic marine debris. *Marine Policy*, **6**, 200–203.

Wilcox C., Mallos N., Leonard G. H., Rodriguez A. and Hardesty B. D. (2016). Using expert elicitation to estimate the impacts of plastic pollution on marine wildlife. *Marine Policy*, **65**, 107–114.

Xanthos D. and Walker T. R. (2017). International policies to reduce plastic marine pollution from single-use plastics (plastic bags and microbeads): a review. *Marine Pollution Bulletin*, **118**, 17–26.

Index

Study Guide

Hrissi K. Karapanagioti and
Ioannis K. Kalavrouziotis
Microplastics in Water and Wastewater,
London, IWA Publishing

Contents

Preface to the Study Guide

This Study Guide has been written to accompany *Microplastics in Water and Wastewater* which was first published by IWA Publishing in 2019. Our intention in preparing the Study Guide has been to facilitate teachers and lecturers using the original book as a textbook and to allow students to study it by themselves (e.g. as part of a distance learning education programme).

Microplastics in Water and Wastewater covers the important subject of microplastics in water and sewage. It starts with the human water cycle and the points where microplastics could interact with water, which are presented in the introductory Chapter 1. Chapter 2 examines indications of the presence of microplastics in freshwater such as in rivers and lakes, as well as dangerous chemicals associated with microplastics in such systems. The following chapters discuss the presence of microplastics in wastewater, investigating their sources (Chapter 3), and their transport through a wastewater treatment plant (WWTP) (Chapter 4). Chapter 5 presents and evaluates sampling methods, processing techniques, and methods of sample analysis that have been used so far to detect microplastics in wastewater. The presence of microplastics in sewage sludge and soils irrigated with sewage or fertilized with mud is assessed in Chapter 6. The microplastic accumulation in waste around the world and their distribution and impact on the environment around the discharge pipes of treated wastewater is examined both with the use of mathematical models (Chapter 7) and field

Editors: Hrissi K. Karapanagioti and Ioannis K. Kalavrouziotis
doi: 10.2166/9781789061697_219

sampling (Chapter 8). Plastic biomedia, presented in Chapter 9, are another source of microplastic originating directly from WWTPs. Possible effects on marine and freshwater microalgae (Chapter 10), soils and agricultural plants (Chapter 11) as well as on human health (Chapter 12) are all evaluated. Finally, Chapter 13 presents the need for strategies that will lead to the reduction of environmental plastic pollution.

This Study Guide has been developed to be closely aligned with the content of the original text. Each Study Guide section includes the title of the relevant chapter, its aims, the expected learning outcomes, key concepts, a study plan, additional literature, and a set of self-assessment exercises and activities. Indicative answers to the self-assessment exercises and activities are provided at the end of the study guide.

Each chapter can be covered within a week and in this way the whole book could be covered in a 13-week semester course. However, selected chapters could equally be combined to form modules that could be used in different courses e.g. Chapters 1, 2, 10, 11, 12, and 13 could be used in a Natural Environment and Pollution or Emerging Pollutants Course; Chapters 3, 4, 5, 6, and 9 could be used in a Wastewater Treatment or Wastewater Management Course; Chapters 7 and 8 in a Marine Pollution Course; and Chapters 1 and 13 in an Environmental Law Course.

The writing team for this Study Guide is made up of some of the original book's chapter authors who have provided material for writing the respective sections related to their chapters, namely: I.K. Kalavrouziotis, H.K. Karapanagioti, K. Katsanou, S. Kordella, D. Korfiatis, M. Leotsinidis, N. Mourgkogiannis, D. Papaioannou, G. Papatheodorou, E. Sazakli, M. Tunali and O. Yenigün. The book editors are also authors of part of the sections corresponding to their own chapters of the textbook. The editors also took care of the consistency of presentation of this Study Guide and ensured completeness where there were any gaps.

Associate Professor Dr Hrissi K. Karapanagioti
Department of Chemistry,
University of Patras

Professor Dr Ioannis K. Kalavrouziotis
Dean of School of Science and Technology,
Hellenic Open University

Patras, 2020

How to use this book

Each section in this Study Guide relates to the chapter of the same number/name in the main textbook.

AIMS

Each section in this Study Guide provides you with a handy summary of what the chapter in the textbook covers.

EXPECTED LEARNING OUTCOMES

Each section also provides you with a list of things you should have learned after reading the relevant textbook chapter.

KEY CONCEPTS

Key concepts are listed as an additional aid for you to focus your attention and monitor your understanding.

STUDY PLAN

The Study Plan table, which appears in each chapter, breaks down the study of that particular chapter into a logical sequence of study periods.

The original textbook sections which are being studied are listed in the table for easy reference.

Relevant items from the Additional Bibliography (which accompanies each chapter) are specified for each Study Period, as well as relevant Evaluation Exercises and Activities.

The Study plan from Chapter 1 is shown below as an example

Study Period	Textbook Sections	Additional Bibliography	Evaluation Exercises/Activity
1st period	1.1–1.3	Papers [1.2], [1.4], [1.6]	S.A. 1.1, Act. 1.1
2nd period	1.4	Paper [1.3], selectively from the Book [1.5]	S.A. 1.2
3rd period	1.5–1.6	Paper [1.1]	S.A. 1.3, S.A. 1.4

SELF-ASSESSMENT EXERCISES

Self-assessment exercises are provided for each chapter. Answers can be found towards the back of this Study Guide.

ACTIVITIES

Suggested activities are provided for each chapter. Indicative answers can be found towards the back of this Study Guide.

CHAPTER 1. PLASTICS AND MICROPLASTICS IN THE HUMAN WATER CYCLE

AIMS

This chapter aims to provide an introduction to the sources of plastics and microplastics and explain how they are introduced by humans as they use water in their everyday life. The features, appearance, and concentrations of plastics and microplastics in different water bodies are presented. The chapter also gives an insight into the already established measures widely taken worldwide to address the problem of their ever-increasing accumulation.

EXPECTED LEARNING OUTCOMES

After studying this chapter, you should have acquired a global view of the new water pollution problem caused by microplastics. More specifically, you should:

- be familiar with the different types of microplastics;
- understand the processes that control the transfer and accumulation of microplastics in various bodies of water;
- be aware of the levels of concentration and the possible effects of microplastics on fresh water and wastewater;
- be able to identify the lack of proper management and the need for measures to be taken to address this pollution problem.

KEY CONCEPTS

Abundance, characteristics, definitions, drinking water, existing legislation, groundwater, seawater, sources, surface water, water cycle.

STUDY PLAN

Study Period	Textbook Sections	Additional Bibliography	Evaluation Exercises/Activity
1st period	1.1–1.3	Papers [1.2], [1.4], [1.6]	S.A. 1.1, Act. 1.1
2nd period	1.4	Paper [1.3], selectively from the Book [1.5]	S.A. 1.2, Act. 1.1
3rd period	1.5–1.6	Chapter [1.1]	S.A. 1.3, S.A. 1.4

Editors: Hrissi K. Karapanagioti and Ioannis K. Kalavrouziotis
doi: 10.2166/9781789061697_0223

The study of this chapter is based on a logical sequence of specific learning topics that can be better studied if they are divided into 3 study periods. The study sequence introduces you first to general issues of defining microplastics and the processes that occur in the anthropogenic water cycle (corresponding to Sections 1.1–1.3 in the main textbook), then to specific issues such as the appearance and transport of microplastics to various water bodies (Section 1.4), and then to the relevant legislation (Sections 1.5–1.6).

Each study period is accompanied by a suggested additional bibliography to be studied if you have more time or if you want to learn more details about one of these topics:

- microplastics and anthropogenic water cycle [1.2, 1.4, 1.6]
- presence and transport of microplastics in different water bodies [1.3, selectively from 1.5]
- relevant legislation [1.1]

At the same time, you should complete the self-assessment exercises and carry out the suggested activity below.

You may also seek either more bibliographic sources through databases or other support materials (studies, videos, case studies, photos, etc.) on the internet. It is recommended, if possible, that you visit various water bodies in your region and try to spot microplastics that are visible to the naked eye and attempt to determine their origin.

ADDITIONAL BIBLIOGRAPHY

[1.1] Brennholt N., Heß M. and Reifferscheid G. (2018). Freshwater Microplastics: Challenges for Regulation and Management. In: Freshwater Microplastics: Emerging Environmental Contaminants?, M. Wagner and S. Lambert (eds.), Springer International Publishing, Heidelberg, pp. 239–272. doi: 10.1007/978-3-319-61615-5.

[1.2] Eerkes-Medrano D., Thompson R. C. and Aldridge D. C. (2015). Microplastics in freshwater systems: a review of emerging threats, identification of knowledge gaps and prioritization of research needs. *Water Research*, **75**, 63–82. doi: 10.1016/j.watres.2015.02.012.

[1.3] Horton A. (2017). A Review of Current Knowledge, Microplastics in the Freshwater Environment. Foundation for Water Research, Marlow, UK. http://www.fwr.org/environw/frr0027.pdf.

[1.4] Koelmans A. A., Hazimah N., Nora M., Hermsen E., Kooi M., Mintenig S. M. and De France J. (2019). Review: Microplastics in freshwaters and drinking water: Critical review and data quality assessment. *Water Research*, **155**, 410–422. doi: 10.1016/j.watres.2019.02.054.

[1.5] Lambert S. and Wagner M. (2018) Microplastics Are Contaminants of Emerging Concern in Freshwater Environments: An Overview. In: Freshwater Microplastics: Emerging Environmental Contaminants?, M. Wagner and S. Lambert (eds.),

Springer International Publishing, Heidelberg, pp. 1–23. doi: 10.1007/978-3-319-61615-5.

[1.6] Rios Mendoza L. M., Karapanagioti H. K. and Ramírez Alvarez N. (2018). Micro (nanoplastics) in the marine environment: Current knowledge and gaps. *Current Opinion in Environmental Science & Health*, **1**, 47–51. doi: 10.1016/j.coesh.2017.11.004.

SELF-ASSESSMENT EXERCISES

S.A. 1.1

What percentage of microplastics originate from the land and what from the ocean? Name some of the land-based sources.

S.A. 1.2

Compare microplastics found in tap water and in bottled water in terms of quantity and quality.

S.A. 1.3

Name some countries that have laws related to Microbead-Free Water and describe what these laws are about.

S.A. 1.4

Answer the following questions

1. The amount of plastic waste in the ocean will continue to grow unless preventive measures are taken.
   True or false?
2. Microplastics and plastics are a threat to:
   a. Fauna and flora
   b. Ecosystems
   c. Humans
   d. All of the above
3. A significant part of marine pollution comes from:
   a. Rivers
   b. Groundwater
   c. Rainfall
   d. Air
4. To avoid increasing the amount of plastics, we must:
   a. Identify their origin
   b. Control their use
   c. Introduce legislation
   d. All of the above
5. What is the main source of microplastics?
   a. Transport of cargo
   b. Oil refining
   c. Airplane emissions
   d. Human activities in general

ACTIVITY

Act. 1.1

Try answering the questionnaire found at the following link: https://www.surveymonkey.com/r/VBBSLF9

CHAPTER 2. ASSOCIATION OF HAZARDOUS COMPOUNDS WITH MICROPLASTICS IN FRESHWATER ECOSYSTEMS

AIMS

This chapter aims to provide an introduction to the hazardous compounds that can be found in microplastics that have been sampled in freshwater systems. It classifies these hazardous compounds under different general categories and explains how they were associated with the microplastics. The present studies related to microplastics in freshwater and the reasons for limited studies in this area are discussed.

EXPECTED LEARNING OUTCOMES

After studying this chapter, you should have an idea of the different hazardous compounds that can be found in the different microplastics. More specifically, you should:

- be familiar with the general categories of hazardous compounds found in microplastics;
- have an understanding of the reasons why the microplastics are associated with the various hazardous compounds;
- have an idea of the concentration ranges of different compounds in freshwater microplastics;
- be able to discuss possible implications due to the association of microplastics with hazardous compounds.

KEY CONCEPTS

Additives, bacteria, concentration, lakes, metals, persistent organic pollutants (POPs), pharmaceuticals, sorption, toxic compounds, viruses.

STUDY PLAN

Study Period	Textbook Sections	Additional Bibliography	Evaluation Exercises/Activity
1st period	2.1–2.2.4	Paper [2.1], selectively from the book [2.2]	S.A. 2.1, S.A. 2.2, Act. 2.1
2nd period	2.2.5–2.4	Paper [2.1], selectively from the book [2.2]	S.A. 2.3, S.A. 2.4

The study of this chapter is based on a logical sequence of specific learning topics that can be better studied if they are divided into 2 study periods. The study sequence introduces you first to the hazardous compounds that are commonly studied in the environment and are associated with microplastics (Sections 2.1–2.2.4), and then to additive compounds that specifically originate from the production of plastics, as well as a discussion of possible implications from the above association (Sections 2.2.5–2.4).

Each study period is accompanied by a suggested additional bibliography to be studied if you have more time or if you want to learn more details about one of these topics:

- environmental pollutants [2.1, 2.2]
- additives [2.1, 2.2]

At the same time, you should proceed with the self-assessment exercises and the suggested activity below.

You may also seek either more bibliographic sources through databases or more support materials (studies, videos, case studies, photos, etc.) on the internet. It is recommended, if possible, to visit a local plastic producer and find out the additives that are used to produce different plastic items.

ADDITIONAL BIBLIOGRAPHY

[2.1] Rios Mendoza L. M., Taniguchi S. and Karapanagioti H. K. (2017). Advanced analytical techniques for assessing the chemical compounds related to microplastics. *Comprehensive Analytical Chemistry*, **75**, 209–240. doi: 10.1016/bs.coac.2016.11.001

[2.2] Takada H. and Karapanagioti H. K. (2019). Hazardous Chemicals Associated with Plastics in the Marine Environment. The Handbook of Environmental Chemistry **78**. Springer International Publishing.

SELF-ASSESSMENT EXERCISES

S.A. 2.1

What do the following abbreviations mean: POPs, PCBs, OCPs, DDT and PAHs?

S.A. 2.2
Explain why microplastics act as vectors of POPs.
S.A. 2.3
What do the following abbreviations mean: BPA, BFR, PFR, PFAS and PBDEs?
S.A. 2.4
What are some of the characteristics that additives can give to final plastic products?

ACTIVITY

Act. 2.1
Compare concentration range values for toxic compounds in microplastics from freshwater environments found in Table 2.1 with corresponding values for toxic compounds in microplastics from marine environments found in Table 2 from paper [2.1] in the additional bibliography. How do they compare?

CHAPTER 3. MICROPLASTICS IN WASTEWATER TREATMENT PLANTS: A LITERATURE REVIEW OF SAMPLING METHODS AND RESULTS

AIMS

This chapter aims to give a comprehensive picture of the existing literature on the presence and sampling of microplastics in wastewater treatment plants (WWTPs) worldwide. In addition, the chapter aims to evaluate the most commonly used sampling techniques, and to present typical concentrations and types of microplastics that can be located at the influent or the effluent of a WWTP. It also briefly presents an emerging type of microplastics, namely pharmaceutical microplastics, that have not being studied in detail in the literature.

EXPECTED LEARNING OUTCOMES

After studying this chapter, you should have a complete picture of the presence of microplastics in WWTPs worldwide. More specifically, you should be able to:

- classify the microplastics found in WWTPs into categories
- evaluate and describe the steps and sampling methods commonly used
- explain the importance of the presence of microplastics in WWTPs
- recognize pharmaceutical microplastics and separate them from other categories of microplastics.

KEY CONCEPTS

Continuous wastewater flow, equivalent population, filters, microplastics, pharmaceutical microplastics, sampling instruments, sampling methods, sampling stages, treatment level, wastewater treatment plants (WWTPs).

STUDY PLAN

Study Period	Textbook Sections	Additional Bibliography	Evaluation Exercises/Activity
1st period	3.1–3.2	Report [3.1], Chapter [3.2]	S.A. 3.1
2nd period	3.2.1	Paper [3.3], Report [3.5]	S.A. 3.2, S.A. 3.3
3rd period	3.2.2	Papers [3.4], [3.6]	Act. 3.1, Act. 3.2, S.A. 3.4
4th period	3.3–3.4	Paper [3.7]	S.A. 3.5, Act. 3.3

The study of this chapter is based on a logical sequence of specific learning topics that can be better studied if they are divided into 4 study periods. The study sequence first introduces you to a generalized knowledge of the presence of microplastics in WWTPs globally, based on the existing bibliography (Sections 3.1, 3.2). Next, a deeper understanding is gained by addressing the steps and sampling methods used for microplastic detection in WWTPs (Section 3.2.1), before proceeding with the study of the most abundant microplastics found in WWTP and the amounts escaping into aquatic ecosystems (Section 3.2.2). Finally, a case study is presented on microplastics found in WWTPs in Greece, according to the literature (Section 3.3).

Each study period is accompanied by a suggested additional bibliography:

- microplastics in WWTPs [3.1, 3.2]
- stages and sampling methods [3.4, 3.5]
- types and quantities of microplastics [3.3, 3.6]
- case study: the situation in WWTPs in Greece [3.7]

At the same time, you should complete the self-assessment exercises and carry out the suggested activities below.

You may also seek either more bibliographic sources through databases or more support materials (studies, videos, case studies, photos, etc.) on the internet. It is recommended, if possible, that you visit a WWTP and, while on site, try to detect plastics and microplastics that are visible to the naked eye at the effluent outlet or in the chlorination tank.

ADDITIONAL BIBLIOGRAPHY

[3.1] GESAMP. (2015). Sources, Fate and Effects of Microplastics in the Marine Environment: A Global Assessment. IMO, London. Retrieved September 28, 2018, from file:///C:/Users/User/Downloads/reports-and-studies-no-90-en.pdf.

[3.2] Karapanagioti H. K. (2017). Microplastics and synthetic fibers in treated wastewater and sludge. In: Wastewater and Biosolids Management, I. K. Kalavrouziotis (ed.), IWA Publishing, London, pp. 77–88.

[3.3] Dyachenko A., Mitchell J. and Arsem N. (2017). Extraction and identification of microplastic particles from secondary wastewater treatment plant (WWTP) effluent. *Analytical Methods*, **9**, 1412–1418.

[3.4] Murphy F., Ewins C., Carbonnier F. and Quinn B. (2016). Wastewater Treatment Works (WwTW) as a Source of Microplastics in the Aquatic Environment. *Environmental Science & Technology*, **50**(11), 5800–8.

[3.5] Magnusson K. and Noren F. (2014) Screening of microplastic particles in and down-stream a wastewater treatment plant, IVL-report C55, https: //www. diva-portal.org/smash/get/diva2:773505/FULLTEXT01.pdf.

[3.6] Ziajahromi S., Neale P. A., Rintoul L. and Leeusch F. D. (2017). Wastewater treatment plants as a pathway for microplastics: Development of a new approach to sample wastewater-based microplastics. *Water Research*, **112**, 93–99.

[3.7] Mourgkogiannis N., Kalavrouziotis I. K. and Karapanagioti H. K. (2018). Questionnaire-based survey to managers of 101 wastewater treatment plants in Greece confirms their potential as plastic marine litter sources. *Marine Pollution Bulletin*, **133**, 822–827. doi: 10.1016/j. marpolbul.2018.06.044.

SELF-ASSESSMENT EXERCISES

S.A. 3.1

At a secondary wastewater treatment plant, 0.005 microplastics per cubic centimeter of wastewater enter the plant in one day. Within, the facility, the removal efficiency is 24% of the initial microplastic loading per treatment stage. If the wastewater supply is $15 \cdot 10^8$ cm^3/d, calculate:

a. the microplastics exiting from each treatment stage (except the final stage) of the WWTP in one day;
b. the microplastics escaping this WWTP in one day;
c. the microplastics escaping the WWTP in one year (365 days).

Assume that the treatment stages are: (1) screening; (2) primary sedimentation tank; (3) aeration tank; and (4) secondary sedimentation tank. The wastewater flow at all stages is constant and equal to $15 \cdot 10^8$ cm^3/d.

S.A. 3.2

Assume that two different wastewater treatment plants, plants A and B, serve the same equivalent population and 5 microplastics per L were introduced in each facility. WWTP A has a flow rate of 1000 m^3/d and a 98% microplastic removal is achieved and WWTP B has a flow rate of 1100 m^3/d and a 99% microplastic removal is achieved. To determine the particle size of the escaping microplastics, the effluent flow of both WWTPs passed through a steel sieve with pore diameter of 0.1 cm. For WWTP A, 20% of the total escaping microplastics (MPs) have size <1 mm, and for WWTP B, 75% of the total escaping MPs have size >1 mm.

How many microplastics escape from each facility in one day? How many microplastics with size >1 mm escape from each facility?

Assume there is no organic matter in the effluents.

S.A. 3.3

Based on the literature, make a list of the instruments/devices used to perform continuous and random sampling. What are the advantages and disadvantages of each method?

S.A. 3.4

A municipal WWTP serves an equivalent population of 10,000 individuals and 4 microplastics per L enter the plant. If the number of escaping microplastics is 560,000 per week, calculate:

a. the removal efficiency of the WWTP;
b. how many synthetic fibers and microbeads are escaping in one day?

Assume that there are 3 microbeads and 1 synthetic fiber per L.

S.A. 3.5

Study Section 3.3 of the book and describe what a pharmaceutical microplastic is. Are pharmaceutical microplastics considered a threat to aquatic ecosystems? Explain.

ACTIVITIES

Act. 3.1

Visit a WWTP in your area. Observe the solids collected in the outer collection bin of the secondary settling tank and locate pharmaceutical microplastics and other plastic categories, and watch out for potential plastic and other solid particles escaping with the wastewater. Categorize the plastic particles you observed and suggest measures to limit their occurrence in aquatic ecosystems.

Act. 3.2

Visit a WWTP in your area. Use a steel bucket and collect a 50 L bulk sample from the chlorination tank. Pass the sample through a sieve (with a diameter of 0.5 mm) and locate plastic that has been collected. Categorize plastic particles based on their diameter and count microplastic particles. Calculate how many microplastics escape from the WWTP in one day.

Act. 3.3

Visit a beach in your area which is located at a point near the discharge pipe of a WWTP. Look for pharmaceutical microplastics that potentially escape the WWTP. Try to find out if the WWTP receives wastewater from a medical unit (hospital, health centre, etc.).

CHAPTER 4. MICROPLASTICS: TRANSPORT AND REMOVAL AT WASTEWATER TREATMENT PLANTS

AIMS

This chapter aims to present an overview of the processes that contribute to solids removal during wastewater treatment and evaluate the effectiveness of these existing solid removal processes to remove microplastics from wastewater. In other words, the reader is offered the opportunity to look inside the heart of the operating system of a WWTP, to understand the fate of microplastics at each treatment stage and see whether their density affects their retention capacity at each stage.

EXPECTED LEARNING OUTCOMES

After studying this chapter, you should understand the contribution of the conventional processes (mechanical, chemical and biological) for solids removal used for each of the individual stages of processing for the retention of microplastics in a WWTP, and for the prevention of their escape to receiving water bodies. More specifically, you should be able to:

- evaluate the retention of solids particles in the pre-processing and primary processing stages;
- have a complete picture of the distribution of microplastics, according to their density, in the primary settlement tank of a WWTP;
- understand and explain the presence of microplastics in the process of biological treatment with activated sludge;
- evaluate the ability to retain microplastics in tertiary WWTPs (the role of membranes and disinfectant solutions such as NaOCl in the structure of microplastic particles);
- understand and be able to interpret the resistance of microplastics both during exposure to a biofilm and to other degradation factors such as oxidation processes, ultraviolet radiation UV and the like;
- analyze and identify those factors that determine precisely the microplastics holding capacity of a WWTP.

KEY CONCEPTS

Biofilm, buoyancy, density, flotation, gravitational partitioning, gravity separation, lipophilicity, polymers, removal, skimming.

STUDY PLAN

Study Period	Textbook Sections	Additional Bibliography	Evaluation Exercises/Activity
1st period	4.1–4.4	Paper [4.3]	S.A. 4.1, Act. 4.1
2nd period	4.5–4.6	Chapter [4.2]	S.A. 4.2
3rd period	4.7	Chapter 6 of the main textbook	S.A. 4.3
4th period	4.8–4.9	Paper [4.1]	S.A. 4.4

The study of this chapter is based on a logical sequence of specific learning topics that can be better studied if they are divided into 4 study periods. The study sequence first provides you with a generalized knowledge of the removal of solids in the various stages of a WWTP (Sections 4.1–4.4), continuing with an introduction to the effects of chemical and microbiological and other processes that affect polymeric materials within a WWTP (Sections 4.5–4.6), then proceeds to discuss sludge processing (Section 4.7). Finally, it makes an overall assessment of the removal of microplastics from WWTPs (Sections 4.8–4.9).

Each study period is accompanied by a suggested additional bibliography:

- removal of solids in the various stages of a WWTP [4.3]
- chemical, microbiological and other factors that affect the fate of microplastics [4.2]
- sludge and microplastics treatment [Chapter 6 of the main textbook]
- removal of microplastics from WWTPs [4.1]

At the same time, you should proceed with the self-assessment exercises and the suggested activity below.

You may also seek either more bibliographic sources through databases or more support materials (studies, videos, case studies, photos, etc.) on the internet. It is recommended, if possible, that you visit a WWTP and try to detect microplastics that are visible to the naked eye at various stages of treatment and in the respective waste collection bins.

ADDITIONAL BIBLIOGRAPHY

[4.1] Carr S. A., Liu J. and Tesoro A. G. (2016). Transport and fate of microplastic particles in wastewater treatment plants. *Water Research*, **91**, 174–182.

[4.2] Harrison J. P., Hoellein T. J., Sapp M., Tagg A. S., Ju-Nam Y. and Ojeda J. J. (2018). Microplastic-associated biofilms: a comparison of freshwater and marine environments. In: Freshwater Microplastics, M. Wagner and S. Lambert (eds.), Springer International Publishing, Heidelberg, pp. 181–201. doi: 10.1007/978-3-319-61615-5.

[4.3] Murphy F., Ewins C., Carbonnier F. and Quinn B. (2016). Wastewater treatment works (WwTW) as a source of microplastics in the aquatic environment. *Environmental Science & Technology*, **50**, 5800–5808.

SELF-ASSESSMENT EXERCISES

S.A. 4.1

Look carefully at Tables 4.1 and 4.2. What is the minimum sample volume to be obtained from the output of the secondary WWTP so that there is at least one microplastic particle in the sample? How does this volume compare to the volume that should be taken from a tertiary WWTP?

S.A. 4.2

Explain the possible changes that may occur to a microplastic during its course in an EEL. Why do they occur?

S.A. 4.3
Where do the majority of microplastics end up to when sewage sludge is treated?
S.A. 4.4
What do the authors of this chapter suggest should be done to improve WWTPs for the removal of microplastics?

ACTIVITY

Act. 4.1
Look at Figure 4.1 and name the tertiary level treatment processes presented in the figure. Then, check other textbooks to find out what other treatment processes are used for tertiary level treatment. Can you identify which ones may remove microplastics?

CHAPTER 5. METHOD DEVELOPMENT FOR MICROPLASTIC ANALYSIS IN WASTEWATER

AIMS

The aims of this chapter are the presentation and discussion of the main sampling strategies for sewage in the various stages of a WWTP – including collection, treatment and analysis. These strategies are performed based on internationally accepted protocols for the qualitative and quantitative determination of microplastics. Based on current literature, an assessment of microplastics at specific WWTPs is presented. The advantages and disadvantages of the analytical techniques for the detection of microplastics in wastewater are pointed out. Finally, issues that need further research are highlighted, such as the presence of microplastics in the activated sludge process.

EXPECTED LEARNING OUTCOMES

After studying this chapter, you should be familiar with sampling, pretreatment, and classification of microplastics, and know all the important considerations related to these procedures when working in a WWTP. More specifically, you should be able to:

- specify the sampling strategy that you will follow in an installation in order to collect representative samples, the analysis of which will determine the quantities of microplastics that are found in a WWTP;
- understand the preparation of the samples you would receive from a WWTP for the detection of microplastics;
- identify the analytical techniques widely used for the qualitative and quantitative determination of microplastics in WWTPs;
- classify microplastics based on accepted international protocols.

KEY CONCEPTS

Chemical digestion, effluent, enzymatic digestion, microscopy, Nile Red dye, particle analysis, sampling design, sampling technique, spectroscopy, wastewater treatment plant (WWTP).

STUDY PLAN

Study Period	Textbook Sections	Additional Bibliography	Evaluation Exercises/Activity
1st period	5.1–5.2	Report [5.3]	S.A. 5.1, 5.2
2nd period	5.3–5.4	Paper [5.2]	S.A. 5.3, 5.4, 5.5, Act. 5.1
3rd period	5.5–5.7	Paper [5.1]	S.A. 5.6

The study of this chapter is based on a logical sequence of specific learning topics that can be better studied if they are divided into 3 study periods. The study sequence initially introduces you to the available literature and the development of a sampling strategy (Sections 5.1–5.2), then continues with sample preparation and analysis techniques (Sections 5.3–5.4), before finally explaining how to evaluate the quality of the results (Sections 5.5–5.7).

Each study period is accompanied by a suggested item from the additional bibliography below:

- sampling strategy [5.3]
- sample processing and analysis [5.2]
- evaluation of the quality of the results [5.1]

At the same time, you should proceed with the self-assessment exercises and the suggested activity below.

You may also seek more bibliographic sources through databases and more support materials (studies, videos, case studies, photos, etc.) on the internet. It is recommended, if possible, to visit an analytical laboratory with ATR-FTIR, micro-FTIR, Raman, or micro-Raman, and try to understand how these various pieces of analytical equipment operate and the challenges associated with their use for monitoring microplastics in water and wastewater.

ADDITIONAL BIBLIOGRAPHY

[5.1] Dyachenko A., Mitchell J. and Arsem N. (2017). Extraction and identification of microplastic particles from secondary wastewater treatment plant (WWTP) effluent. *Analytical Methods*, **9**, 1412–1418.

[5.2] Maes T., Jessop R., Wellner N., Haupt K. and Mayes A. G. (2017). A rapid-screening approach to detect and quantify microplastics based on fluorescent tagging with Nile Red. *Scientific Reports*, **7**, 44501.

[5.3] Sedlak M., Sutton R., Box C., Sun J. and Lin D. (2017). Sampling and Analysis Plan for Microplastic Monitoring in San Francisco Bay and Adjacent National Marine Sanctuaries. San Francisco Estuary Institute Contribution. No. 819. San Francisco Estuary Institute, Richmond, CA.

SELF-ASSESSMENT EXERCISES

S.A. 5.1

Study Figure 5.1 and explain how sampling time and random sampling affect the concentration of microplastics that will eventually be measured by a researcher.

S.A. 5.2

Usually in a WWTP there is a flow meter after pretreatment. However, if it does not exist or it is damaged or we want to confirm the values it measures, what is a quick way to measure the sewage flow?

S.A. 5.3

List common digestion techniques for removing organic matter from aqueous solutions.

S.A. 5.4

In marine samples, it is common for microplastics to be separated based on density. Why can we not use the same technique for sewage?

S.A. 5.5

Describe the principle of operation of the screening technique of samples with Nile red dye to locate microplastics in a sewage sample taken after the digestion of organic matter.

S.A. 5.6

What are the quality control parameters that should be considered when performing an instrumental analysis for microplastics?

ACTIVITY

Act. 5.1

Make a list of the materials you would need if you were to make your own laboratory method for microplastic monitoring of treated wastewater.

CHAPTER 6. MICROPLASTICS IN SEWAGE SLUDGE: CAPTURED BUT RELEASED?

AIMS

This chapter aims to explain why sewage sludge land application is the main source of microplastics, and especially of synthetic fibers, in the terrestrial environment. The presence of micro- and nanoplastics in sludge and the techniques commonly used for their detection are discussed. The disadvantages of the analytical techniques (based on cost and time) used by the scientific community to quantify micro- and nanoplastics are presented. Particle numbers ending up on agricultural land each year through the deposit of sludge onto land ecosystems as soil fertilizer are calculated.

EXPECTED LEARNING OUTCOMES

After studying this chapter, you should better understand the fate of microplastics after they are removed from wastewater. More specifically, you should be able to:

- describe the transport and incorporation of microplastics into sewage sludge;
- identify and analyze the concentration of microplastics in sewage sludge, based on international literature;
- evaluate and interpret the fate of microplastics in sewage sludge;
- predict the behavior and impact of microplastics on soil when they are incorporated into sewage sludge that is then used as soil fertilizer or conditioner.

KEY CONCEPTS

Agriculture, clarifiers, contamination, effluents, fibers, microplastics, nanoplastics, sewage sludge, sludge reuse, soil conditioner.

STUDY PLAN

Study Period	Textbook Sections	Additional Bibliography	Evaluation Exercises/Activity
1st period	6.1–6.2	Papers [6.1], [6.3]	S.A. 6.1
2nd period	6.3	Report [6.2]	S.A. 6.2
3rd period	6.4–6.5	Paper [6.4], different chapters in the Book [6.5]	S.A. 6.3, Act. 6.1

The study of this chapter is based on a logical sequence of specific learning topics that can be better studied if they are divided into 3 study periods. The study sequence first introduces you to the removal of microplastics from sewage and their transfer to sludge (Sections 6.1–6.2), then continues into the methodologies and results of various studies that have measured microplastics in sludge (Section 6.3), and finally discusses the transfer of microplastics to the environment, depending on the manner of sludge management (Sections 6.4–6.5).

Each study period is accompanied by a suggested additional bibliography:

- transport of microplastics from sewage to sludge [6.1, 6.3]

- measurement of microplastics in sludge [6.2]
- transfer of microplastics from sludge to the environment [6.4, 6.5]

At the same time, you should proceed with the self-assessment exercises and the suggested activity below.

You may also seek either more bibliographic sources through databases or more support materials (studies, videos, case studies, photos, etc.) on the internet. It is recommended, if possible, that you visit a WWTP and try to detect microplastics that are visible to the naked eye in sludge treatment and sludge storage tanks.

ADDITIONAL BIBLIOGRAPHY

[6.1] Carr S. A., Liu J. and Tesoro A. G. (2016). Transport and fate of microplastic particles in wastewater treatment plants. *Water Research*, **91**, 174–182.

[6.2] Lusher A. L., Hurley R. R., Vogelsang C., Nizzetto L. and Olsen M. (2018). Mapping microplastics in sludge – Report for the Norwegian Environment Agency (No. M907). Norwegian Institute for Water Research, Oslo.

[6.3] Murphy F., Ewins C., Carbonnier F. and Quinn B. (2016). Wastewater Treatment Works (WwTW) as a Source of Microplastics in the Aquatic Environment. *Environmental Science & Technology*, **50**, 5800–5808.

[6.4] Nizzetto L., Futter M. and Langaas S. (2016). Are Agricultural Soils Dumps for Microplastics of Urban Origin? *Environmental Science & Technology*, **50**, 10777–10779.

[6.5] Takada H and Karapanagioti HK (2019) Hazardous Chemicals Associated with Plastics in the Marine Environment. The Handbook of Environmental Chemistry **78**. Springer International Publishing.

SELF-ASSESSMENT EXERCISES

S.A. 6.1

Study Table 6.1. What are the most common types of microplastics found in sewage sludge? Which polymers are they made of?

S.A. 6.2

Why is it difficult to compare the results of various studies with each other?

S.A. 6.3

What percentage of sewage sludge is used for soil application, and what is the amount of microplastics that are estimated to be added to the soil each year by this practice in the European Union?

ACTIVITY

Act. 6.1

Search the internet for ways to reduce the release of microplastic fibers when washing clothes.

CHAPTER 7. MODELING MICROPLASTICS TRANSPORT AND FATE IN THE MARINE ENVIRONMENT AROUND A WASTEWATER EFFLUENT DISCHARGE PIPE

AIMS

The aims of this chapter are the presentation of the transformation processes of microplastics, as well as presentation of the basic principles governing their sedimentation and horizontal transport in the marine environment. The chapter shows that microplastics from wastewater treatment plants are an important source of sea pollution.

EXPECTED LEARNING OUTCOMES

After studying this chapter, you should have a complete picture of the behavior of microplastic particles in the marine environment. More specifically, you should be able to:

- explain the transformation processes of microplastics.
- describe the advection–diffusion equation of microplastics and explain the meaning of each term.
- solve simple applications of the above equation.

KEY CONCEPTS

Advection, aggregation, degradation, diffusion, equations, flow, marine environment, mathematical model, sedimentation, transport.

STUDY PLAN

Study Period	Textbook Sections	Additional Bibliography	Evaluation Exercises/Activity
1st period	7.1–7.2	Paper [7.1], [7.2], [7.3]	Act. 7.1
2nd period	7.3–7.5	Paper [7.4] Any introductory book on fluid mechanics. Any introductory book on ordinary differential equations.	S.A. 7.1, 7.2, 7.3

The study of this chapter is based on a logical sequence of specific learning topics that can be better studied if they are divided into 2 study periods. The study sequence first introduces you to transformation processes of microplastics (Sections 7.1–7.2), and then continues to present transport processes and mathematical modelling (Section 7.3–7.5).

Each study period is accompanied by a suggested additional bibliography:

- transformation processes [7.1, 7.2, 7.3]
- transport processes and mathematical model [7.4]

At the same time, you should proceed with the self-assessment exercises and the suggested activity below.

You may also seek either more bibliographic sources through databases or more support materials (studies, videos, case studies, photos, etc.) on the internet.

ADDITIONAL BIBLIOGRAPHY

[7.1] Mourgkogiannis N., Kalavrouziotis I. K. and Karapanagioti H. K. (2018). Questionnairebased survey to managers of 101 wastewater treatment plants in Greece confirms their potential as plastic marine litter sources. *Marine Pollution Bulletin*, **133**, 822–827.

[7.2] Besseling E., Quik J. T. K., Sun M. and Koelmans A. A. (2017). Fate of nano- and microplastic in freshwater systems: a modeling study. *Environment Pollution*, **220**, 540–548.

[7.3] Quik J. T. K., Van de Meent D. and Koelmans A. A. (2014). Simplifying modeling of nanoparticle aggregation-sedimentation behavior in environmental systems: a theoretical analysis. *Water Research*, **62**, 193–201.

[7.4] Arvidsson R., Molander S., Sandén B. A. and Hassellöv M. (2011). Challenges in exposure modeling of nanoparticles in aquatic environments. *Human and Ecological Risk Assessment*, **17**, 245–262.

SELF-ASSESSMENT EXERCISES

S.A. 7.1

Solve the differential Equation (7.9) and prove that its solution is Equation (7.10).

S.A. 7.2

Prove the relationship described by Equation (7.14) following the steps outlined in the text.

S.A. 7.3

Prove the relationship described by Equation (7.16) following the steps outlined in the text.

ACTIVITY

Act. 7.1

Access the article: Andrady A. L. (2011), Microplastics in the marine environment. *Marine Pollution Bulletin* 62, 1596–1605. Read Section 1.3 of the article and write a paragraph on the processes of degradation of microplastics in the marine environment.

CHAPTER 8. EVALUATING WASTEWATER EFFLUENT AS A SOURCE OF MICROPLASTICS IN ENVIRONMENTAL SAMPLES

AIM

The aim of this chapter is to examine the uncertainty related to locating the source of plastics in environmental samples and especially samples taken from outside of WWTP effluent discharge points. It proposes a framework that can be used in a multiple line of evidence approach to assess the source of microplastics in the environment.

EXPECTED LEARNING OUTCOMES

After studying this chapter, you should be familiar with the plastic items that are indicators of marine contamination from wastewater effluents. More specifically, you should be able to:

- discuss taxonomic evidence related to microplastics' origin;
- outline microplastic indicators;
- outline wastewater indicators;
- discuss issues related to indicator correlation, mass balance and standardization.

KEY CONCEPTS

Anthropogenic litter, correlation, indicators, marine debris, mass balance, mismanaged waste, plastic pollution, source allocation, standardization, taxonomic evidence.

STUDY PLAN

Study Period	Textbook Sections	Additional Bibliography	Evaluation Exercises/Activity
1st period	8.1–8.2	Paper [8.1] and Chapter 4 of the main textbook	S.A. 8.1
2nd period	8.3–8.4	Papers [8.2], [8.3]	S.A. 8.2
3rd period	8.5–8.8	Paper [8.4]	Act. 8.1

The study of this chapter is based on a logical sequence of specific learning topics that can be better studied if they are divided into 3 study periods. The study sequence introduces you first to anecdotal evidence of marine debris sources, which is the method of source evaluation with the highest uncertainty (Sections 8.1–8.2), and then to taxonomic evidence with plastic indicators as well as wastewater indicators (Sections 8.3–8.4), and then to higher certainty source evaluation techniques such as correlation and mass balance (Sections 8.5–8.8).

Each study period is accompanied by a suggested additional bibliography:

- anecdotal evidence [8.1]
- taxonomic evidence and indicators [8.2, 8.3]
- correlation and mass balance [8.4]

At the same time, you should proceed with the self-assessment exercises and the suggested activity below.

You may also seek either more bibliographic sources through databases or more support materials (studies, videos, case studies, photos, etc.) on the internet. It is recommended, if possible, that you visit a receiving water close to a WWTP and try to detect macroplastic debris that can be associated with wastewater discharge.

ADDITIONAL BIBLIOGRAPHY

[8.1] Carr S. A., Liu J. and Tesoro A. G. (2016). Transport and fate of microplastic particles in wastewater treatment plants. *Water Research*, **91**, 174–82.

[8.2] Mason S. A., Garneau D., Sutton R., Chu Y., Ehmann K., Barnes J., Fink P., Papazissimos D. and Rogers D. L. (2016). Microplastic pollution is widely detected in US municipal wastewater treatment plant effluent. *Environ Pollut.*, **218**, 1045–54.

[8.3] McCormick A., Hoellein T. J., Mason S. A., Schluep J. and Kelly J. J. (2014). Microplastic is an abundant and distinct microbial habitat in an urban river. *Environmental Science & Technology*, **48**(20), 11863–71.

[8.4] Mintenig S. M., Bäuerlein P. S., Koelmans A. A., Dekker S. C. and van Wezel A. P. (2018). Closing the gap between small and smaller: towards a framework to analyse nano- and microplastics in aqueous environmental samples. *Environmental Science: Nano*, **5**(7), 1640–9.

SELF-ASSESSMENT EXERCISES

S.A. 8.1

Name the different source evaluation techniques and put them in order of decreasing uncertainty.

S.A. 8.2

Even though microbeads and pellets have a similar shape (they are both spherical), they cannot be mistaken for each other. Explain why.

ACTIVITY

Act. 8.1

What are some of the densities of different polymers that can help explain Figure 8.7?

CHAPTER 9. POLLUTION OF BEACHES AND WATERCOURSES BY PLASTIC BIOMEDIA

AIM

The aim of this chapter is to describe the problem of environmental pollution due to the loss of plastic biomedia used in WWTPs. Different pollution cases and different WWTPs incidents are presented and discussed to determine the parameters that can lead to such problems.

EXPECTED LEARNING OUTCOMES

After studying this chapter, you should be able to describe the whole life cycle of plastic biomedia as well as plastic pollution caused by their use. More specifically, you should be able to:

- explain how plastic biomedia are used in a WWTP;
- outline who are the main users of plastic biomedia;
- describe different pollution cases involving plastic biomedia release to the environment.

KEY CONCEPTS

Beach cleanup, biocarriers, biological treatment, incidents, land-based sources, monitoring, moving bed biofilm reactor (MBBR), participatory science, system malfunctions, waterways.

STUDY PLAN

Study Period	Textbook Sections	Additional Bibliography	Evaluation Exercises/Activity
1st period	9.1–9.3	Papers [9.2], [9.3]	Act. 9.1, S.A. 9.1
2nd period	9.4–9.6	Report [9.1]	Act. 9.2
3rd period	9.7–9.8	Report [9.1]	S.A. 9.2

The study of this chapter is based on a logical sequence of specific learning topics that can be better studied if they are divided into 3 study periods. The study sequence introduces you first to some general information related to the use of plastic biomedia (Sections 9.1–9.3), and then discusses how plastic biomedia are spread and outlines different pollution incidents (Sections 9.4–9.6), before finally presenting different system malfunctions (Sections 9.7–9.8).

Each study period is accompanied by a suggested additional bibliography:

- plastic biomedia in wastewater purification [9.2, 9.3]
- plastic biomedia in the natural environment [9.1]
- system malfunctions [9.1]

At the same time, you should complete the self-assessment exercises and carry out the suggested activities below.

You may also seek either more bibliographic sources through databases or more support materials (studies, videos, case studies, photos, etc.) on the internet. It is recommended, if possible, that you visit a WWTP that uses a fluidized bed bioreactor with plastic biomedia.

ADDITIONAL BIBLIOGRAPHY

[9.1] Bencivengo P., Barreau C., Bailly C. and Verdet F. (2018). Sewage Filter Media and Pollution of the Aquatic Environment. Surfrider Foundation Europe Report. Water Quality and Marine Litter programme, Biarritz, France. See: https://www.surfrider.eu/wp-content/uploads/2018/08/biomedia-pollution-report.zip.

[9.2] Sfaelou S., Papadimitriou C. A., Manariotis I. D., Rouse J. D., Vakros J. and Karapanagioti H. K. (2016). Treatment of low-strength municipal wastewater containing phenanthrene using activated sludge and biofilm process. *Desalination and Water Treatment*, **57**, 12047–12057. DOI: 10.1080/19443994.2015.1048735.

[9.3] Sfaelou S., Karapanagioti H. K. and Vakros J. (2015). Studying the formation of biofilms on supports with different polarity and their efficiency to treat wastewater. *Journal of Chemistry*, No 734384.

SELF-ASSESSMENT EXERCISES

S.A. 9.1

What are the advantages and the disadvantages of using a fluidized bed bioreactor with plastic biomedia for wastewater treatment?

S.A. 9.2

List possible system malfunctions that can lead to spills of plastic biomedia.

ACTIVITIES

Act. 9.1

Try to find WWTPs in your area that are using plastic biomedia for water treatment.

Act. 9.2

Check the website of a local supplier of biomedia in your country. Check out the types of plastic biomedia. Visit a beach in your area and try to spot these plastic biomedia.

CHAPTER 10. EFFECTS OF MICROPLASTICS ON FRESHWATER AND MARINE MICROALGAE

AIMS

This chapter provides an overview of the effects of different types of microplastics, such as polystyrene and polyethylene, on the functions of different types of microalgae. Also, it presents the effects of the concentration of microplastics, their size and surface load, as well as contact time, on the growth of algae, chlorophyll concentration and photosynthetic yield.

EXPECTED LEARNING OUTCOMES

After studying this chapter, you should have extensive knowledge of the results of research studies that focus on the interaction of microplastics and microalgae. More specifically, you should be able to:

- understand how microplastics affect the functions of microalgae;
- be familiar with the range of selected parameter values related to microalgae (concentration, size, etc.) that has been studied in the literature;
- identify gaps in the literature and develop further research topics.

KEY CONCEPTS

Algae type, algal growth, chlorophyll concentration, combined effects, concentration, exposure time, microalgae, microplastic size, microplastic type, photosynthetic efficiency.

STUDY PLAN

Study Period	Textbook Sections	Additional Bibliography	Evaluation Exercises/Activity
1st period	10.1	Paper [10.1]	S.A.10.1, Act 10.1
2nd period	10.2.1	Paper [10.3]	Act. 10.2
3rd period	10.2.2–10.3	Paper [10.2]	S.A. 10.2, Act. 10.3, S.A. 10.3

The study of this chapter is based on a logical sequence of specific learning topics that can be better studied if they are divided into 3 study periods. The study sequence introduces you first to general issues concerning the global problem of plastic pollution (Section 10.1), and then to specific issues such as the effects of microplastics on the development of algae (Section 10.2.1), before addressing other effects and providing conclusions (Sections 10.2.2–10.3).

Each study period is accompanied by a suggested additional bibliography:

- the global problem of plastic pollution [10.1]
- the effects of microplastics on algae development [10.3]
- other implications and conclusions [10.2]

At the same time, you should complete the self-assessment exercises and carry out the suggested activities below.

You may also seek either more bibliographic sources through databases or more support materials (studies, videos, case studies, photos, etc.) on the internet.

ADDITIONAL BIBLIOGRAPHY

[10.1] Cole M., Lindeque P., Halsband C. and Galloway T. S. (2011). Microplastics as contaminants in the marine environment: A review. *Marine Pollution Bulletin*, **62** (12), 2588–2597. doi: 10.1016/j.marpolbul.2011.09.025.

[10.2] Prata J. C., Lavorante B. R. B. O., Montenegro M. and da Guilhermino C. B. S. M. L. (2018). Influence of microplastics on the toxicity of pharmaceuticals procainamide and doxycycline on the marine microalgae Tetraselmischuii. *Aquatic Toxicology*, **197**, 143–152. doi: 10.1016/j.aquatox.2018.02.015.

[10.3] Zhang C., Chen X., Wang J. and Tan L. (2017). Toxic effects of microplastic on marine microalgae Skeletonemacostatum: Interactions between microplastic and algae. *Environmental Pollution*, **220**, 1282–1288. doi: 10.1016/j.envpol.2016.11.005.

SELF-ASSESSMENT EXERCISES

S.A. 10.1

In two sentences, describe the global problem of marine pollution from plastics. What are the major types of plastic produced?

S.A. 10.2

Describe how microplastics affect photosynthetic performance and chlorophyll concentration.

S.A. 10.3

What are the limits of the studies that have been done so far on the effects of microplastics on microalgae and what should future research issues be?

ACTIVITIES

Act. 10.1

Give an example of a plastic that you use regularly in your daily life and determine its type.

Act. 10.2

Explain how you would measure algal growth.

Act. 10.3

What other pollutants can be detected in combination with microplastics? Why?

CHAPTER 11. POSSIBLE EFFECTS ON PLANTS DUE TO MICROPLASTICS IN SOILS FROM WASTEWATER EFFLUENT REUSE OR SEWAGE SLUDGE APPLICATION

AIMS

This chapter presents the environmental fate of microplastics in agro-ecosystems, with reference to the complex interactions between physicochemical and biological soil factors and the properties of the microplastics themselves. The chapter aims to study the sources of microplastics in agro-ecosystems, the mechanisms and behavior of microplastics in soil, and their reaction with organisms and plants.

EXPECTED LEARNING OUTCOME

After studying this chapter, you should have an idea of the general problems associated with the accumulation of microplastics in soils due to wastewater reuse. More specifically, you should be able to:

- determine microplastic sources in agricultural soils;
- estimate the quantities of microplastics added to agricultural soils by applying wastewater and sludge from biological wastewater treatment;
- understand the degradation parameters of microplastics in soils, and the properties of the additives that they carry;
- estimate the effects and risks of the presence of microplastics in soils.

KEY CONCEPTS

Additives, agriculture, degradation, irrigation, plant, quantities, sewage sludge, soil, soil pollution, wastewater reuse.

STUDY PLAN

Study Period	Textbook Sections	Additional Bibliography	Evaluation Exercises/Activity
1st period	11.1	Papers [11.1], [11.2]	–
2nd period	11.2	Papers [11.3], [11.4]	S.A. 11.1, Act. 11.1
3rd period	11.3	Papers [11.5], [11.6]	S.A. 11.2
4th period	11.4–11.5	Paper [11.7]	S.A.11.3, Act. 11.2

The study of this chapter is based on a logical sequence of specific learning topics that can be better studied if they are divided into 4 study periods. The study sequence initially introduces you to the subject of the presence of microplastics after application of wastewater and sludge to agricultural soils (Section 11.1). Then, based on literature data, it presents calculations of microplastics that accumulate in agricultural soils after application of by-products from sewage biological treatment and, at the same time, the load of microplastics in crops is also calculated (Section 11.2). The degradation of microplastics in the ground is then studied, while making a reference to the most important additives found in plastics (Section 11.3). Finally, the effects of microplastics on agricultural soils (soil ecosystems, crops and animals) is studied (Sections 11.4–11.5).

Each study period is accompanied by a suggested additional bibliography:

- microplastics and the application of waste onto agricultural lands [11.1, 11.2]
- microplastics load, depending on the crop [11.3, 11.4]
- decomposition of microplastics and additional chemicals [11.5, 11.6, 11.7]
- possible effects of microplastics on agricultural land [11.8]

At the same time, you should complete the self-assessment exercises and carry out the suggested activities below.

You may also seek either more bibliographic sources through databases or more support materials (studies, videos, case studies, photos, etc.) on the internet. It is recommended, if possible, that you visit different agricultural lands and crops, and try to detect microplastics that are visible to the naked eye.

ADDITIONAL BIBLIOGRAPHY

[11.1] Bläsing M. and Amelung W. (2018). Plastics in soil: Analytical methods and possible sources. *Science of the Total Environment*, **612**, 422–435. doi: 10.1016/j.scitotenv.2017.08.086.

[11.2] Ng E. L., Huerta Lwanga E., Eldridge S. M., Johnston P., Hu H. W., Geissen V. and Chen D. (2018). An overview of microplastic and nanoplastic pollution in agroecosystems. *Science of the Total Environment*, **627**, 1377–1388. doi: 10.1016/j.scitotenv.2018.01.341.

[11.3] Nizzetto L., Futter M. and Langaas S. (2016) Are agricultural soils dumps for microplastics of urban origin? *Science of the Total Environment*, **50**(20), 10777–10779. doi.org/10.1021/acs.est.6b04140.

[11.4] Mahon A. M., O'Connell B., Healy M. G., O'Connor I., Officer R., Nash R. and Morrison L. (2017). Microplastics in sewage sludge: effects of treatment. *Environmental Science & Technology*, **51**(2), 810–818. doi.org/10.1021/acs.est.6b04048.

[11.5] Teuten E. L., Saquing J. M., Knappe D.R. U., Barlaz M. A., Jonsson S., Bjorn A., Rowland S. J., Thompson R. C., Galloway T. S., Yamashita R., Ochi D., Watanuki Y., Moore C., Viet P. H., Tana T. S., Prudente M., Boonyatumanond R., Zakaria M. P., Akkhavong K., Ogata Y., Hirai H., Iwasa I., Mizukawa K., Hagino Y.,

Imamura A., Saha M. and Takada H. (2009). Transport and release of chemicals from plastics to the environment and to wildlife. *Philosophical Transactions of the Royal Society B: Biological Sciences*, **364**(1526), 2027–2045. doi: 10.1098/rstb.2008.0284.

[11.6] Soares A., Guieysse B., Jefferson B., Cartmell E. and Lestera J. N. (2008). Nonylphenol in the environment: A critical review on occurrence, fate, toxicity and treatment in wastewaters. *Environment International*, **34**(7), 1033–1049. doi: 10.1016/j.envint.2008.01.004.

[11.7] Sun J., Wu X. and Gan J. (2015). Uptake and metabolism of phthalate esters by edible plants. *Environmental Science & Technology*, **49**(14), 8471–8478. doi: 10.1021/acs.est.5b01233.

SELF-ASSESSMENT EXERCISES

S.A. 11.1

According to the studies presented in this chapter, what are the values of the concentrations of microplastics both in treated wastewater and in sewage sludge from WWTPs using biological treatment.

S.A. 11.2

What are the main chemical additives in plastics and what are the possible consequences for organisms?

S.A. 11.3

What factors can affect the presence of microplastics in soil?

ACTIVITIES

Act. 11.1

Assuming an annual cultivated area of 1 Ha (10,000 m^2), for a crop of your choice, use information contained in this chapter together with your own research to calculate the possible amounts of microplastics that could be added to the soil by applying treated wastewater for irrigation.

Act. 11.2

Draw the various paths of microplastics through soil, using 'soil', 'plants', 'microorganisms' and 'worms' in the soil, 'animals', etc., as reference points.

CHAPTER 12. POSSIBLE EFFECTS OF MICROPLASTICS ON HUMAN HEALTH

AIMS

The aims of this chapter are to provide training and expertise on the potential health effects of microplastics, and to present potentially dangerous substances and particles, as well as examine the major exposure routes to humans. The effects of exposure to micro- and nanoplastic particles through the three main exposure paths are examined separately. Potentially adverse effects from exposure to chemicals released by microplastics and acting as endocrine disruptors are described. Information is provided from recent biomonitoring studies as well as epidemiological surveillance around the world. Special reference is made to two main representatives of these chemicals: bisphenol A and phthalates. Finally, methods of pathogen transportation and entry into the human body through microplastics are examined.

EXPECTED LEARNING OUTCOMES

After studying this chapter, you should be able to better understand information related to how the possible effects of microplastics on human health are determined. More specifically, you should be able to:

- understand the processes and mechanisms through which microplastics adversely affect human health, particularly through:
  - particle exposure
  - exposure to chemicals (and especially to endocrine disruptors), and
  - exposure to pathogenic microorganisms;
- understand the way endocrine disruptors work and to learn about the effects of their actions and how they differ in men and women;
- understand the importance of "critical developmental windows" for exposure to endocrine disruptors;
- understand the concept of the interaction of the environment and of pollutants with human health;
- understand how guidelines are drawn up and health decisions are taken by using the results of toxicological and epidemiological studies.

KEY CONCEPTS

Biomonitoring studies, Bisphenol A, dermal route of exposure, endocrine disruptors, inhalation, masculinization, nanoparticles, oral route of exposure, phthalates, reproductive toxicity.

STUDY PLAN

Study Period	Textbook Sections	Additional Bibliography	Evaluation Exercises/Activity
1st period	12.1–12.2.1	Paper [12.5]	S.A. 12.1, Act. 12.1,12.2
2nd period	12.2.2–12.2.2.1	Papers [12.1], [12.4]	S.A. 12.2
3rd period	12.2.2.2–12.3	Papers [12.2], [12.3]	S.A. 12.3, 12.4, Act. 12.3

The study of this chapter is based on a logical sequence of specific learning topics that can be better studied if they are divided into 3 study periods. The study sequence first introduces you to the possible effects due to exposure to microplastic particles (Sections 12.1–12.2.1), and then those due to exposure to chemicals such as bisphenol A (Sections 12.2.2–12.2.2.1), and finally, those due to exposure to phthalates and other additives as well as exposure to microorganisms (Sections 12.2.2.2–12.3).

Each study period is accompanied by a suggested additional bibliography:

- particle exposure [12.5]
- exposure to bisphenol A [12.1, 12.4]
- exposure to phthalates, additives, microorganisms [12.2, 12.3]

At the same time, you should complete the self-assessment exercises and carry out the suggested activities below.

You may also seek either more bibliographic sources through databases or more support materials (studies, videos, case studies, photos, etc.) on the internet.

ADDITIONAL BIBLIOGRAPHY

[12.1] Fenichel P., Chevalier N. and Brucker-Davis F. (2013). Bisphenol A: an endocrine and metabolic disruptor. *Annales d Endocrinologie*, **74**(3), 211–220. doi: 10.1016/j.ando.2013.04.002

[12.2] Meeker J. D., Sathyanarayana S. and Swan S. H. (2009). Phthalates and other additives in plastics: human exposure and associated health outcomes. *Philosophical Transactions of the Royal Society B: Biological Sciences*, **364**, 2097–2113. doi: 10.1098/rstb.2008.0268

[12.3] Revel M., Châtel A. and Mouneyrac C. (2018). Micro (nano) plastics: A threat to human health? *Current Opinion in Environmental Science and Health*, **1**, 17–23. doi: 10.1016/j.coesh.2017.10.003

[12.4] Rochester J. R. (2013). Bisphenol A. and human health: a review of the literature. *Reproductive Toxicology*, **42**, 132–155. doi: 10.1016/j.reprotox.2013.08.008

[12.5] Wright S. L. and Kelly F. J. (2017). Plastic and Human Health: A Micro Issue? *Environmental Science & Technology*, **51**(12), 6634–6647. doi.org/10.1021/acs.est.7b00423

SELF-ASSESSMENT EXERCISES

S.A. 12.1

Outline the pathways for micro- and nanoplastic particles entering the body. What are the target organs and what are the health effects?

S.A. 12.2

Briefly (in about 100 words) describe how endocrine disruptors disrupt the normal functioning of the endocrine system.

S.A. 12.3

Describe the main effects of phthalates on the various stages of human development.

S.A. 12.4

How can microplastics transmit pathogenic microorganisms into the human body? What are the possible adverse effects?

ACTIVITIES

Act. 12.1

Consider 4 activities and habits in everyday life that involve the use of plastic products and suggest the use of alternative products.

Act. 12.2

Make a note of the number next to the recycling symbol on various plastics (numbers 1 to 7). Report their importance and explain their differences in terms of ease of recycling convenience.

Act. 12.3

Prepare a questionnaire of up to 10 questions (or look for one on the internet) and ask your family, friends or colleagues, etc., the amount of plastic products used daily, the types of plastics most used, the extent of recycling and reuse of plastic products, and whether participants are aware of the potential effects of plastics on their health.

CHAPTER 13. THE NEED FOR A GLOBAL PLASTIC STRATEGY

AIMS

This chapter aims to briefly present the global, growing problem of plastic pollution in freshwater and in the world's oceans. It also presents the most important global and cross-border policies, strategies and initiatives for the prevention and reduction of plastic waste in aqueous ecosystems. The basic characteristics of an effective policy/strategy are reviewed.

EXPECTED LEARNING OUTCOMES

After studying this chapter, you should have a good idea of the regulations and policies that are in effect to battle plastic pollution. More specifically, you should be able to:

- discuss the seriousness of the problem of contamination of aquatic ecosystems with plastics;
- outline the key components of major global and cross-border policies, strategies and initiatives related to the prevention and reduction of waste in the aquatic environment;

- determine the key features of a successful policy/strategy for the prevention and reduction of plastic waste in the aquatic environment.

You should also have enriched your skills with the ability:

1. to evaluate the effectiveness of a plastics policy/strategy;
2. to participate in policymaking or in designing initiatives to prevent and reduce plastic pollution in aquatic ecosystems;
3. to raise public awareness about the pollution of aquatic ecosystems with plastic waste.

Changes are also expected regarding your attitude towards the problem. More specifically, you should be able to embed environmentally friendly practices into your everyday life, having become more aware of issues related to plastic pollution.

KEY CONCEPTS

Framework directives, initiatives, marine litter, plastic litter, plastic pollution, plastic waste, policy, single-use plastics, strategies, waste, water.

STUDY PLAN

Study Period	Textbook Sections	Additional Bibliography	Evaluation Exercises/Activity
1st period	13.1–13.2.1	Papers [13.2], [13.3]	S.A. 13.1, 13.2
2nd period	13.2.2–13.3	European Directive [13.1]	S.A.13.3, Act. 13.1

The study of this chapter is based on a logical sequence of specific learning topics that can be better studied if they are divided into 2 study periods. The study sequence introduces you first to general issues of plastic pollution and global strategies and policies on the subject (Sections 13.1–13.2.1), and then to European Directives and Strategies, as well as to other local initiatives (Sections 13.2.2–13.3).

Each study period is accompanied by a suggested additional bibliography as follows:

- global strategies and policies [13.2]
- European Directives and Strategies, and other local initiatives [13.1]

At the same time, you should proceed with the self-assessment exercises and the suggested activity below.

You may also seek either more bibliographic sources through databases or more support materials (studies, videos, case studies, photos, etc.) on the internet.

ADDITIONAL BIBLIOGRAPHY

[13.1] European Commission (2018). A European Strategy for Plastics in a Circular Economy. Communication from the Commission to the European Parliament, the Council, the European Economic and Social Committee and the Committee of the Regions. COM (2018) 28 final; 16 January 2018. European Commission, Brussels. See: https://eur-lex.europa.eu/legal-content/EN/TXT/HTML/?uri=CELEX:52018DC0028&from=EN (accessed 5 May 2020).

[13.2] Laina E. (2018). Expert group identifies barriers and response options on marine litter and microplastics. *Environmental Policy and Law*, **48**, 168–171.

[13.3] Xanthos D. and Walker T. R. (2017). International policies to reduce plastic marine pollution from single-use plastics (plastic bags and microbeads): a review. *Marine Pollution Bulletin*, **118**, 17–26. doi: 10.1016/j.marpolbul.2017.02.048

SELF-ASSESSMENT EXERCISES

S.A. 13.1

Give examples of global strategies that have been involved in preventing and/or reducing plastic pollution.

S.A. 13.2

Study paper [13.2] in the additional bibliography. What is the main challenge dealing with the microplastics pollution problem?

S.A. 13.3

What was the goal of the LIFE DEBAG project?

ACTIVITY

Act. 13.1

Visit the LIFE DEBAG project website and provide examples of project activities.

ANSWERS TO SELF-ASSESSMENT EXERCISES

S.A. 1.1

The sources of microplastics are 80% land-based and 20% ocean-based. Land-based sources include personal care products, air-blasting processes, improperly disposed plastics, leachate from landfills, etc.

S.A. 1.2

Studies found twice as many microplastics in bottled water than in tap water. Fragments from packaging and/or the bottling process were predominant in bottled water but other types of microplastic materials were also found. Microplastics in tap water were mainly fibers.

S.A. 1.3

Countries that have passed laws banning microbeads in cosmetics and personal health products include The Netherlands, Canada, Ireland and the UK. These laws aim to reduce water pollution from microbeads entering the water bodies through treated sewage.

S.A. 1.4

The correct answers are:

1. True
2. (d) All of the above
3. (a) Rivers
4. (d) All of the above
5. (d) Human activities in general

S.A. 2.1

Check Section 2.1.

POPs: Persistent organic pollutants
PCBs: Polychlorinated biphenyls
OCPs: Organochlorine pesticides
DDT: Dichlorodiphenyltrichloroethane
PAHs: Polycyclic aromatic hydrocarbons

S.A. 2.2

POPs are persistent organic compounds that are hydrophobic and are found in polluted areas. Since microplastics have a lipophilic surface, they can sorb POPs that are in the sea or in fresh water in a polluted area, acting the same way as the suspended organic matter that also sorbs POPs from the water phase.

S.A. 2.3

Check Section 2.5.

BPA: Bisphenol A
BFRs: Brominated flame retardants
PFRs: Phosphorus flame retardants
PFAS: Perfluoroalkylated compounds
PBDEs: Polybrominated diphenyl ethers

S.A. 2.4

Some examples of the characteristics that additives can give to final plastic products are color, resistance to heat and resistance to aging.

S.A. 3.1

a. The microplastic removal efficiency of the WWTP at each stage is as follows:
   - *Microplastics allowed to escape from the first treatment stage*: these are the initial 100% minus the removal efficiency (which is 24% for each treatment stage) = 76% = 0.76

     0.76 × 0.005 microplastics/m^3 = 0.76 × 5 microplastics/L = 3.8 microplastics/L, i.e. 3.8 × 1500 m$^{3/}$d = 3.8 × 1,500,000 L/d = 5,700,000 microplastics/d
   - *Microplastics allowed to escape from the second treatment stage:* these are the previous 76% minus the removal efficiency (which is 24% for each treatment stage) = 52% = 0.52

     0.52 × 5 = 2.6 microplastics/L, 2.6 × 1500 m$^{3/}$d = 2.6 × 1,500,000 L/d = 3,900,000 microplastics/d
   - *Microplastics allowed to escape from the third treatment stage:* these are the previous 52% minus the removal efficiency (which is 24% for each treatment stage) = 28% = 0.28

     0.28 × 5 = 1.4 microplastics/L, 1.4 × 1500 m$^{3/}$d = 1.4 × 1,500,000 L/d = 2,100,000 microplastics/d

b. The microplastics escaping this WWTP in one day are as follows:
   - these are the previous 28% (found in (a) above) minus the removal efficiency (which is 24% for each treatment stage) = 4% = 0.04

     0.04 × 5 = 0.2 microplastics/L, 0.2 × 1500 m$^{3/}$d = 0.2 × 1,500,000 L/d = 300,000 microplastics/d

     The overall removal efficiency for microplastics at this WWTP is 96%. It can be calculated in either of the following 2 ways:

   1. 4 treatment stages × 24% removal efficiency per treatment stage = 96%; or
   2. 5 microplastics/L and 1,500,000 L/d = 7,500,000 microplastics/d entering the WWTP, and 300,000 microplastics/d escaping the WWTP, giving:

      (7,500,000 – 300,000)/7,500,000 = 96%.

c. The microplastics escaping the WWTP in one year (365 days) are as follows:

   - 300,000 microplastics/d × 365 d/year = 109,500,000 microplastics/year (So, although the removal efficiency for microplastics is 96%, the WWTP releases approximately 110 million microplastics into aquatic bodies every year).

S.A. 3.2

WWTP A: Since the removal efficiency is 98%, the escaping percentage is $100 - 98 = 2\% = 0.02$

WWTP B: Since the removal efficiency is 99%, the escaping percentage is $100 - 99 = 1\% = 0.01$

The following calculations can be used to calculate the number of microplastics escaping from each facility in one day given that 5 microplastics/L were introduced in each facility:

WWTP A: $0.02 \times 5 = 0.1$ microplastics/L; flow rate 1,000 m^3/d = 1,000,000 L/d

Therefore, 0.1 microplastics/L × 1,000,000 L/d = 100,000 microplastics/d

Since 20% of the total escaping microplastics (MPs) have size <1 mm, the rest (80%) will have size >1 mm. Thus, escaped microplastics with size >1 mm for WWTP A = 100,000 microplastics/d × 0.8 = 80,000 microplastics/d

WWTP B: $0.01 \times 5 = 0.05$ microplastics/L; flow rate 1,100 m^3/d = 1,100,000 L/d

Therefore, 0.05 microplastics/L × 1,100,000 L/d = 55,000 microplastics/d

Escaped microplastics with size >1 mm for WWTP B = 55,000 microplastics/d × 0.75 = 37,500 microplastics/d

S.A. 3.3

In continuous sampling, the electrical device used is mainly pumps that ensure continuous sewage flow.

In random sampling, mainly steel buckets (not plastic ones) are used to prevent contamination of the sample, and a specific volume of sewage is usually obtained. However, a pump can also be used until the required volume is obtained.

Continuous sampling		Random sampling	
Advantages	**Disadvantages**	**Advantages**	**Disadvantages**
Continuous effluent flow	Filters are clogged due to high organic concentration of influents	Cost-effective	Non representative samples
Comprehensive overview of the experimental procedure	Cost-intensive	Flexibility in the sampling field	

S.A. 3.4

Assume that the microplastics that end up with the sewage are synthetic fibers and microbeads. Assume that an average resident consumes 200 L/person/d.

a. Therefore, a population equivalent of 10,000 individuals corresponds to a flow rate of 10,000 × 200 = 2,000,000 L/d = 2,000 m^3/d.

There are 560,000 microplastics/week → 80,000 microplastics/d escaping the plant.

If (1 – z) is the removal efficiency, then z is the percentage escaping = 80,000 (microplastics/d)/(4 microplastics/L × 2,000,000 L/d) →z = 0.01

So, the removal efficiency of this WWTP is (1 – 0.01) = 0.99 = 99%.

b. There are (1/4) × 80,000 = 20,000 synthetic fibers/day and (3/4) × 80,000 = 60,000 microbeads/day escaping from this WWTP.

S.A. 3.5

Pharmaceutical microplastics are defined as microplastics with a diameter of 1 nm to 5 mm which are by-products of medical-pharmaceutical use (e.g. pills).

Yes, since concerns are not limited to the amounts of pharmaceutical microplastics that escape receiving water bodies but also to the pharmaceutical compounds that remain on the plastics.

S.A. 4.1

Based on Table 4.1, we see that there are 373 microplastic particles in 111,787 gallons. Thus, 373 microplastics/111,787 gallons = 0.0033 microplastics/gallon = 1 microplastic per 300 gallons (using the rule of three).

Based on Table 4.2, in the tertiary WWTP a sample of 50,898 gallons is required to measure one microplastic, i.e. a volume 170 times greater than the minimum volume (300 gallons) that would be required in a secondary WWTP. Therefore, we can say with certainty that the sampling that took place was sufficient and that the tertiary treatment has significantly better performance in removing microplastics from the effluent.

S.A. 4.2

Since plastics are resistant to chemical and microbiological factors and their residence time in a WWTP is relatively short, their alteration during their residence in a WWTP is not expected.

S.A. 4.3

The majority of microplastics removed from a WWTP are in the final product, i.e. in the biosolids/sewage sludge.

S.A. 4.4

The authors believe that the efficiency of the removal is excellent. Any new technology would make a very small increase in the performance that would not be worth the cost and effort that would be caused in the WWTP by the integration of such a new technology that has not yet solved all operating problems.

S.A. 5.1

It seems that there is a tendency to measure lower microplastic concentrations with increasing sampling time. In other words, random sampling and short sampling time can lead to inaccurate results.

S.A. 5.2

One method is to take a container of at least 10 L and measure the time it takes to fill it with sewage in the outlet pipe, taking care that there are no losses. The sampling and measurement should be repeated at least 3 times and the average calculated.

S.A. 5.3

It can be digested with chemicals e.g. Fenton reagents, hydrogen peroxide, potassium hydroxyl or using enzymes e.g. cytoplasm, protease.

S.A. 5.4

The problem is that there are other solids in wastewater that have a similar density to microplastics.

S.A. 5.5

The microplastics are transferred by filtration from the sample to the surface of a black filter made of a material that should not be polymeric. The surface is moistened with a Nile red dye solution in acetone so that the solvent evaporates quickly. The filter is illuminated in blue and is studied with a stereoscope that has an orange filter. The microplastics that have been wetted with the dye fluoresce while the filter does not.

S.A. 5.6

See all the quality control parameters in Table 5.3. Examples include a lab blank, a field blank, etc.

S.A. 6.1

The most common microplastics found in sewage sludge are polyethylene and polyester fibers.

S.A. 6.2

A comparison of the various studies made is difficult as each WWTP is different. They may differ in terms of the equivalent population, the sources of sewage, or the processes used to treat the sewage or sewage sludge, etc.

S.A. 6.3

In the European Union, about 40% of sewage sludge is used for agricultural purposes. It is estimated that about 43,000–630,000 tons of microplastics are added to European arable land each year through the application of sewage sludge.

S.A. 7.1

The characteristic polynomial is: $D\lambda^2 - v\lambda - K = 0$.

Roots: $\lambda_1 = \dfrac{-\sqrt{v^2 + 4DK} + v}{2D}$ and $\lambda_2 = \dfrac{\sqrt{v^2 + 4DK} + v}{2D}$

The general solution of Equation (7.9) is: $n = Ae^{\lambda_1 x} + Be^{\lambda_2 x}$

Boundary conditions:

$n = 0$ as $x \to \infty$ is satisfied if $B = 0$

$n(x = 0) = n_0 \Rightarrow A = n_0$

Finally, the solution of Equation (7.9) is Equation (7.10) with $\lambda = -\lambda_1$

S.A. 7.2

$$m\frac{dv}{dt} = -F_D \Rightarrow m\frac{dv}{dt} = -\frac{6A\mu}{a}v \Rightarrow \frac{dv}{v} = -\frac{6A\mu}{ma}dt$$

$$\Rightarrow \ln v = -\frac{6A\mu}{ma}t + C \Rightarrow v = v_0 e^{-\frac{6A\mu}{ma}t} \quad (v = v_0 \text{ for } t = 0)$$

$$x = \int v dt + C_1 \Rightarrow x = -v_0\frac{ma}{6A\mu}e^{-\frac{6A\mu}{ma}t} + C_1$$

$$\Rightarrow x = v_0\frac{ma}{6A\mu}\left(1 - e^{-\frac{6A\mu}{ma}t}\right) \quad (x = 0 \text{ for } t = 0)$$

$$\Rightarrow x = v_0\frac{ma}{6A\mu}\left(1 - \frac{v}{v_0}\right) \Rightarrow v = v_0 - \frac{6A\mu}{ma}x$$

S.A. 7.3

$$m\frac{dv}{dt} = -F_D \Rightarrow m\frac{dv}{dt} = -Cv^2 \Rightarrow \frac{dv}{v^2} = -\frac{C}{m}dt$$

$$\Rightarrow -\frac{1}{v} = -\frac{C}{m}t + C_2 \Rightarrow v = \frac{1}{\frac{C}{m}t - C_2} \Rightarrow v = \frac{1}{\frac{C}{m}t + \frac{1}{v_0}} \quad (v = v_0 \text{ for } t = 0)$$

$$x = \int v dt + C_3 \Rightarrow x = \frac{m}{C}\ln\left(\frac{C}{m}t + \frac{1}{v_0}\right) - \frac{m}{C}\ln\left(\frac{1}{v_0}\right) \quad (x = 0 \text{ for } t = 0)$$

$$\Rightarrow x = \frac{m}{C}\ln\left(\frac{1}{v}\right) - \frac{m}{C}\ln\left(\frac{1}{v_0}\right) \Rightarrow x = \frac{m}{C}\ln\left(\frac{v_0}{v}\right) \Rightarrow v = v_0 e^{-\frac{C}{m}x}$$

S.A. 8.1

The source evaluation techniques in order of decreasing uncertainty are: mass balance, correlation, wastewater indicators, taxonomic evidence and anecdotal evidence.

S.A. 8.2

Although microbeads and pellets seem to overlap in their spherical shape definition, they differ in size. Microbeads are smaller than 1 mm, whereas the diameter of plastic pellets ranges between 1 and 5 mm.

S.A. 9.1

The answer is provided in Sections 9.2.3.2 and 9.2.3.3 as well as in the additional bibliography, e.g. an advantage is the use of the existing aeration tanks.

S.A. 9.2

The answer is provided in Section 9.7.2, e.g. failure of safety systems or limitations of a combined sewer system.

S.A. 10.1

Plastic waste can be found all over the world from the Arctic to the Mediterranean Sea and the Pacific Ocean. It is estimated that 8 million tons of plastic are released into the oceans each year.

The types of microplastics commonly found include polyvinyl chloride (PVC), nylon, polyethylene (PE), polypropylene (PP), polystyrene (PS) and polyethylene terephthalate (PET), which together represent 95% of the global plastic production.

S.A. 10.2

In general, a decrease in chlorophyll concentration and photosynthetic activity was observed. Microplastics can have different effects depending on the types of algae present in the aquatic system and the concentration of the microplastics.

S.A. 10.3

The types of plastics and types of algae tested should be expanded to fill the gaps in the literature.

Greater emphasis should be placed on the combined effects of microplastics with other materials to stimulate the natural environment (e.g. by changing interactions and relationships) and to quickly understand the complexity of the problem.

S.A. 11.1

See Section 11.2.2 of the main textbook, e.g. 10,000 microplastics per m^3 could be taken as the mean value for treated wastewater and thousands of microplastics per kg is common for sludge.

S.A. 11.2

See Section 11.3.1 of the main textbook, e.g. Bisphenol A could affect growth of crops.

S.A. 11.3

See Section 11.4 of the main textbook, e.g. agricultural activities such as plowing.

S.A. 12.1

Micro- and nanoplastics can enter the body through inhalation, ingestion, and transdermal absorption.

After inhalation, particles, depending on their size, are retained by the upper respiratory tract (>1 μm) and will undergo mucociliary clearance or descend to the lower respiratory system (<1 μm), where they can settle and cause pneumonia, induce the production of reactive oxygen species (ROS), or lead to secondary gene toxicity, with possible effects on the development of fibrosis and/or cancer.

After ingestion, microparticles enter the gastrointestinal tract and can pass into the lymphatic system where they infect the liver, spleen, kidneys and heart. In addition, nanoparticles can penetrate the blood-brain barrier, and can also pass via the placenta into fetal circulation.

Penetration through the stratum corneum of the skin is limited to particles ≤ 100 nm, so only nanoparticles are expected to penetrate via this route.

S.A. 12.2

Some chemicals mimic a natural hormone and bind to its receptor, thereby producing either an excessive response to a stimulus (e.g. a growth hormone resulting in increased muscle mass), or a response at an inappropriate time (e.g., insulin production when not necessary). Other endocrine disrupters bind to receptors and prevent the normal activity of the hormone (e.g. the developmental hormone that is required for normal development). Also, there are others that directly stimulate or inhibit the endocrine system and cause overproduction or underproduction of hormones (e.g. hyper- or hypo-activity of thyroid).

S.A. 12.3

During the prenatal and first infancy period, phthalates can cause increased prevalence (i.e. existing cases) of hypospadias, cryptorchidism and testicular cancer in newborn boys, as well as a significant reduction in anogenital distance and undescended testes.

In males, they also lead to delayed onset of adolescence and masculinization, as well as quantitative and qualitative reduction of semen.

In females, they can lead to early breast development (premature thelarche) and to early pubertal onset.

In both sexes, phthalates affect the cardiovascular system (increased risk of coronary disease, atherosclerosis, increased diastolic blood pressure), the respiratory system (allergies, rhinitis, asthmatic reactions and inflammation of the lungs), thyroid function (hypothyroidism and hyperthyroidism), mental development (serious neurological diseases, such as deafness, mental retardation, and cerebral dysfunction, with concomitant mobility disorders).

S.A. 12.4

Microplastic surfaces are ideal for the attachment of bacteria, colonization and biofilm development. Biofilms remain on the surface of various types of plastics such as polyethylene (PE) and terephthalate polyethylene (PET). Given their durability, microorganisms can be transferred directly to human tissue (gastrointestinal tract or lung) and alter the normal microbial tissue community. In this way, defense mechanisms can be bypassed, resulting in infections and other immune responses, especially in weakened tissues that have already been affected by micro- and nanoplastic particles.

S.A. 13.1

See Section 13.2.1, e.g. the Honolulu Strategy.

S.A. 13.2

It seems that this problem is very complex, since there are so many different aspects such as political, economical, technological, etc., that have to be taken into consideration for its solution and, at the same time, many sectors are affected by its implications.

S.A. 13.3

See Section 13.2.2. The ultimate goal of the LIFE DEBAG project was to reduce plastic bags in the marine environment of Greece, both locally (in Syros Island, the project's pilot area) and nationally.

INDICATIVE ANSWERS TO ACTIVITIES

Act. 1.1

If you have any questions (e.g. What is the size of the microplastics?), read the relevant information again in Chapter 1.

When you have finished studying all the chapters in the main textbook, visit the questionnaire again and see if your answers have changed.

Act. 2.1

The units of the concentrations and the magnitude of the values are similar.

Act. 3.1

Chapter 3 describes the stages where plastics detected with the naked eye were found in WWTPs in Greece.

Pharmaceutical microplastics collected from WWTPs.

Act. 3.2

Chapter 3 describes how to take samples in the various treatment stages of a WWTP.

Sampling with a steel bucket in a chlorination tank.

Act. 3.3

Pharmaceutical microplastics found on a beach located near the effluent pipe of a WWTP.

Act. 4.1

According to Figure 4.1, tertiary processes are sand filtration and chlorination. Examples of additional tertiary processes include nutrient removal, activated carbon sorption, advanced oxidation processes, and membrane filtration. Any filtration process – but especially membrane filtration – can remove fine particles such as microplastics.

Act. 5.1

You would need:

- Nile Red dye
- solvent/acetone
- black non-polymeric material filter
- filtration device
- stereoscope with orange filter
- blue lamp
- black box

Act. 6.1

Various technological solutions can be found, such as the fiber traps or filters found in washing machines, but there are other ideas related to prevention.

Act. 7.1

Degradation caused by the action of light is called photodegradation. For common polymers such as LDPE, HDPE, PP, and nylons found in the marine environment photodegradation is the main degradation mechanism. The UV-B radiation in sunlight initiates photo-oxidative degradation on the plastic surface. Once initiated, the degradation can also proceed as a slow oxidative breakdown at moderate temperatures (called thermooxidative degradation) for some time without the need for further exposure to UV radiation. On degradation, oxygen-rich functional groups are generated on the plastic surface.

The other types of degradation processes, such as biodegradation, are orders of magnitude slower compared to light-induced oxidation. However, although photodegradation initiated by solar UV radiation is a very efficient mechanism in plastics exposed in air or lying on a beach surface, when the same plastic material is exposed to sunlight at the same location but while floating in seawater, degradation is considerably slower.

Act. 8.1

Unknown to most people, many polymers have extreme density values. However, even the common ones have differences in densities e.g. the density of polystyrene foam is 0.05 g/cm^3 whereas the density of PET is 1.38 g/cm^3 or of polytetrafluoroethylene (PTFE) is 2.2 g/cm^3.

Act. 9.1

Each country has a different approach to wastewater treatment. However, many industrial WWTPs use plastic biomedia.

Act. 9.2

Here is an example picture found from a simple Google search using the word "plasticmedia".

It is not easy to find a single example of biomedia on any beach unless there has been a recent local accident. Here is a picture of an example of biomedia found on a pebble beach in Ithaka, Greece, in June 2020, possibly originating from an aquaculture wastewater treatment unit.

Act. 10.1
E.g. Plastic bottles; generally, PET.
Act. 10.2
Cell counting, spectrophotometric methods, flow cytometer, biomass weight measurement (dry).
Act. 10.3
See Section 10.2.3.2, e.g. pharmaceutical compounds, metals.
Act. 11.1
See Section 11.2 in the textbook and similarly calculate the microplastics for your selected crop. For example: for a potato crop, the annual irrigation rate is 3500 m^3/ha. The mean value of microplastic particles/m^3 of treated wastewater is 10,000. Finally, the annual microplastic load in soil after potato cultivation irrigated with treated wastewater is 3500 × 10,000 = 35 million microplastics/ha.
Act. 11.2

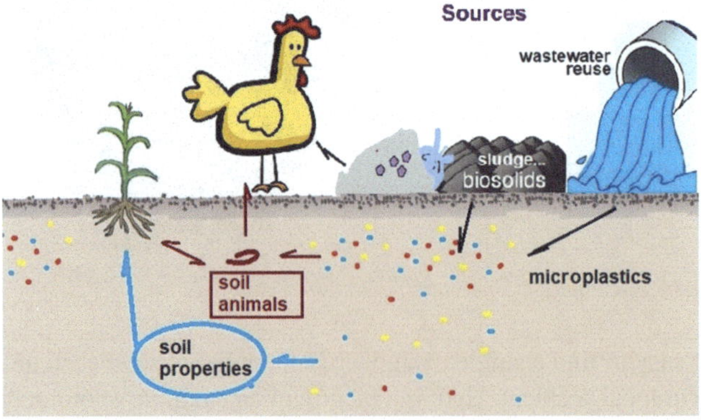

Act. 12.1

A. Storage of food in plastic containers.
Alternative: Buy and store food in glass or metal containers. Use glass drinking water bottles.

B. Heating food in plastic containers in the microwave.
Alternative: Heating food on porcelain dishes.

C. Use of plastic toys by children.
Alternative: Wooden, rubber and fabric toys.

D. Use of plastic clothes, bags and furniture.
Alternative: Use of natural materials (natural fibers, cotton, wood, etc.).

Act. 12.2

Plastics are classified into 7 main categories, depending on the type of synthetic polymer from which they are made. These types differ in their ease of recycling and are:

Type 1: Polyethylene terephthalate (PET) – Easily recyclable
Type 2: High Density Polyethylene (HDPE) – Easily recyclable
Type 3: Polyvinyl chloride (PVC) – Very difficult to recycle
Type 4: Low Density Polyethylene (LDPE) – Manageable
Type 5: Polypropylene (PP) – Manageable
Type 6: Polystyrene (PS) – Hardly recyclable
Type 7: Polycarbonate and others – Very difficult to recycle

Act. 12.3

The questionnaire must be anonymous. Get information on demographics (age, gender, education, income, class). Questions should not overlap. Questions should be of a closed type with multiple choices.

Act. 13.1

The project website is http://www.lifedebag.eu/.

The project was an integrated campaign for the prevention and reduction of plastic bags in the marine environment that addressed all stakeholders differently. It included information activities for consumers (e.g. via a door to door campaign and questionnaires), stakeholder engagement through meetings and door to door education activities for students of all levels. Example of activities included in the campaign are:

- "Plastic Bag Free Weeks";
- Information workshops and events;
- Forum meetings with national stakeholders (supermarkets, plastic industry, etc.) through which policy recommendations were drafted and submitted to the Greek Parliament, contributing to the national legislation concerning plastic bag reduction.

IWA Publishing's authorised EU representative for General Product Safety Regulations is Diane D'Arras, 15 rue Duret, 75116 Paris, France, e-mail: safety@iwap.co.uk.

Printed and bound by CPI Group (UK) Ltd, Croydon, CR0 4YY
16/06/2026
02140886-0001